T0074264

Advances in Intelligent Systems and Computing

Volume 720

About this Series

The series "Advances in Intelligent Systems and Computing" contains publications on theory, applications, and design methods of Intelligent Systems and Intelligent Computing. Virtually all disciplines such as engineering, natural sciences, computer and information science, ICT, economics, business, e-commerce, environment, healthcare, life science are covered. The list of topics spans all the areas of modern intelligent systems and computing.

The publications within "Advances in Intelligent Systems and Computing" are primarily textbooks and proceedings of important conferences, symposia and congresses. They cover significant recent developments in the field, both of a foundational and applicable character. An important characteristic feature of the series is the short publication time and world-wide distribution. This permits a rapid and broad dissemination of research results.

More information about this series at http://www.springer.com/series/11156

Wojciech P. Hunek · Szczepan Paszkiel
Editors

Biomedical Engineering and Neuroscience

Proceedings of the 3rd International Scientific Conference on Brain-Computer Interfaces, BCI 2018, March 13–14, Opole, Poland

Editors
Wojciech P. Hunek
Faculty of Electrical Engineering, Automatic Control and Informatics
Opole University of Technology
Opole
Poland

Szczepan Paszkiel
Faculty of Electrical Engineering, Automatic Control and Informatics
Opole University of Technology
Opole
Poland

ISSN 2194-5357 ISSN 2194-5365 (electronic)
Advances in Intelligent Systems and Computing
ISBN 978-3-319-75024-8 ISBN 978-3-319-75025-5 (eBook)
https://doi.org/10.1007/978-3-319-75025-5

Library of Congress Control Number: 2018931881

Printed on acid-free paper

This Springer imprint is published by the registered company Springer International Publishing AG part of Springer Nature
The registered company address is: Gewerbestrasse 11, 6330 Cham, Switzerland

Preface

With the next issue of the volume 'Advances in Intelligent Systems and Computing,' we are happy to present the proceedings of the 3rd International Scientific Conference BCI. The event was held at Opole University of Technology in Poland between 13 and 14 March 2018. Since 2014, the conference has taken place every two years at the Faculty of Electrical Engineering, Automatic Control and Informatics, Opole University of Technology. During the conference, the speakers and presenters focused on the issues regarding new trends in the modern brain–computer interfaces and control engineering, including neurobiology–neurosurgery, cognitive science–bioethics, biophysics–biochemistry, modeling–neuroinformatics, BCI technology, biomedical engineering, control and robotics, computer engineering, and neurorehabilitation–biofeedback.

The previous two BCI conferences brought into focus several important solutions with regard to both scientific and engineering problems. We can also emphasize that the two last events attracted over 1000 followers representing the biggest national academic centers and industrial companies. In 2016, the collection of the papers was published in the form of a scientific monograph entitled 'Contemporary problems in biomedical engineering and neurosciences,' Szczepan Paszkiel and Jan Sadecki, Eds., Opole University of Technology Press, 2016. Apart from the presentation of full papers, the scientific program also included a number of practical demonstrations covering, for example, the online control of mobile robot and unmanned aerial vehicle using the BCI technology. The awards were sponsored by D-Link Corp., which together with the Faculty of Electrical Engineering, Automatic Control and Informatics, Opole University of Technology, also provided financial support to the conference. During two meetings, the experts gave a variety of presentations including two plenary speeches made by Mieczysław Pokorski, Professor of Medical Sciences from Mossakowski Medical Research Centre Polish Academy of Sciences, regarding the issues related to 'Theory of aging of the body' and 'Neural maps of environment: forming of simple memories.'

We can also note that IC BCI 2018 takes place under the honorary auspices of the Minister of Science and Higher Education Republic of Poland; this year Prof. Janusz Kacprzyk from Systems Research Institute of Polish Academy of Sciences was appointed to be the chairman of the scientific committee, while Szczepan Paszkiel, PhD, is the chairman of the organizing committee. Two last BCI meetings gathered a group of outstanding professionals such as Prof. Marcin Czerwiński, Prof. Włodzislaw Duch, Prof. Grzegorz Francuz, Dr. Tomasz Halski, Prof. Michał Kuczyński, Dr. Dariusz Łątka, Prof. Dariusz Man, Rev. Prof. Piotr Morciniec, Prof. Roman Olejnik, Prof. Tadeusz Skubis, Prof. Jan Szczegielniak. This year the conference is overseen by a local organizing committee having the technical skills, and the team consists of the following staff: Natalia Browarska, Eng., Łukasz Debita, Eng., Robert Kania, Eng., Marek Krok, MSc Eng., Piotr Szpulak, MSc Eng., Tomasz Wacławek, Eng., etc.

Wojciech P. Hunek
Szczepan Paszkiel

Organization

The 3rd International Scientific Conference on Brain-Computer Interfaces BCI 2018 is organized by the Faculty of Electrical Engineering, Automatic Control and Informatics, Opole University of Technology.

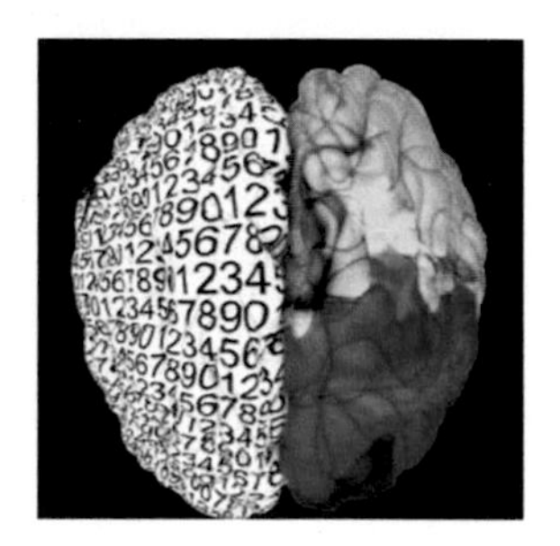

Scientific Committee

Chairman of Scientific Committee

Janusz Kacprzyk	Systems Research Institute Polish Academy of Sciences, Warsaw, Poland

Members of Scientific Committee

Ryszard Beniak	Opole University of Technology, Poland
Tomasz Boczar	Opole University of Technology, Poland
Sebastian Borucki	Opole University of Technology, Poland
Andrzej Cichoń	Opole University of Technology, Poland
Marcin Czerwiński	Institute of Immunology and Experimental Therapy Polish Academy of Sciences, Poland
Włodzisław Duch	Nicolaus Copernicus University in Toruń, Poland

Organizing Committee

Chairman of Organizing Committee

Szczepan Paszkiel

Members of Organizing Committee

Tomasz Bernat
Natalia Browarska
Łukasz Debita
Bartosz Dziurzyński
Robert Kania
Małgorzata Kapica
Marta Korzańska
Marek Krok
Łukasz Mateja
Barbara Miszuda
Piotr Szpulak
Tomasz Wacławek

Ministry of Science
and Higher Education
Republic of Poland
Jarosław Gowin

The Commission of Metrology Polish Academy of Sciences, Branch of Katowice

Adrian Czubak, Opolskie Province Governor

Andrzej Buła, Marshall of the Opolskie Voivodeship

Arkadiusz Wiśniewski, Mayor of Opole City

Michał Siek, Opole Education Superintendent

Prof. Marek Tukiendorf, Rector Opole University of Technology

Contents

Volume Editors

Wojciech P. Hunek obtained PhD and habilitation degrees in Electrical Engineering and Automatic Control and Robotics from the Faculty of Electrical, Control and Computer Engineering, Opole University of Technology, in 2003 and 2012, respectively. He holds the post of Deputy Dean for Education at the Faculty of Electrical, Control and Computer Engineering, Opole University of Technology, and works as Associate Professor at Institute of Control Engineering. He is also the Head of Control Systems and Industrial Automation team. He has authored or co-authored about 90 papers, most of which are concerned with the up-to-date topics in multivariable control and systems theory. He has also reviewed about 145 scientific papers and 105 industrial projects. He is coordinator and participant of several research grants.

Szczepan Paszkiel, PhD Eng. works as an Assistant professor at the Department of Biomedical Engineering at Opole University of Technology. He obtained PhD in Technical Sciences (automatics and robotics) in 2011. His research interests focus on modern methods of control systems, brain–computer technology, and modeling of neuronal cell fractions. Szczepan Paszkiel PhD is the author of more than one hundred publications including those from ISI Master Journal List and a lecturer on several dozen scientific conferences, festivals, and panel discussions. He is the winner of many prizes and competitions including those organized by

Ministry of Science and Higher Education, Republic of Poland, and the recipient of grants for young scientists. He is also the co-editor of the monograph entitled 'Contemporary problems of biomedical engineering and neurosciences,' Opole University of Technology Publishing, 2016; the author of the first handbook in Poland entitled 'Brain–computer Interfaces. Neuroinformatics,' 2014; a member of Metrology Commission, Polish Academy of Sciences, Katowice Branch; an expert of the National Centre for Research and Development, the Centre of OPI Processing Information in the National Research Centre (PIB), an expert of European Commission and Wroclaw Research Centre EIT+; a member of expertise panels and pre-panels in the National Centre for Research and Development including being a chairman of expert teams and a head expert.

Virtual Reality Based Simulators for Neurosurgeons - What We Have and What We Hope to Have in the Nearest Future

Dariusz Latka[1,2(✉)], Marek Waligora[2], Kajetan Latka[1,2,6], Grzegorz Miekisiak[1,5], Michal Adamski[3], Klaudia Kozlowska[3], Miroslaw Latka[3], Katarzyna Fojcik[3], Dariusz Man[4], and Ryszard Olchawa[4]

[1] Department of Neurosurgery, University Hospital in Opole, Opole University, Opole, Poland
dlatka@mp.pl, kajetanlatka@icloud.com, gmiekisiak@gmail.com

[2] Center for Education and Development in Medicine Vital Medic Education, Kluczbork, Poland
mwaligora@vitalmedic.pl

[3] Department of Biomedical Engineering, Faculty of Fundamental Problems of Technology, Wroclaw University of Science and Technology, Wroclaw, Poland
{klaudia.kozlowska,miroslaw.latka}@pwr.edu.pl

[4] Institute of Physics, Opole University, Opole, Poland
dariuszman@vp.pl

[5] Department of Neurosurgery, Regional Medical Center, Polanica-Zdroj, Poland

[6] Specialist District Neurological Center, Opole, Poland

Abstract. High levels of manual skills, good visual-motor coordination, excellent imagination and spatial awareness are the main factors determining the success of neurosurgeons. Proficiency in neurosurgical skills used to be acquired through hands-on training in cadaver labs and in real operating theatres under master neurosurgeon supervision. Most recently, virtual reality (VR) and augmented reality (AR) computer simulations have also been considered as tools for education in the neurosurgical training. The authors review existing solutions and present their own concept of a simulator which could become the useful tool for planning, simulation and training of a specific neurosurgical procedure using patient's imaging data. The benefits of simulator are particularly apparent in the context of neurovascular operations. It is the field in which it is very difficult for young neurosurgeons to gain proficiency because of the lack of experience caused by the competition between microsurgery and endovascular techniques.

Keywords: Virtual reality · Medical training · Neurosurgical simulator

1 Introduction

The rapid development of virtual reality (VR) and augmented reality (AR) over the last few years has created an opportunity for applications in medicine. In the surgical training virtual simulators have become an attractive and increasingly important alternative to the classic cadaver hands-on practical exercises. Until now, they have been widely used in many surgical specialties. An essential part of the surgical residency programs is

W. P. Hunek and S. Paszkiel (Eds.): BCI 2018, AISC 720, pp. 1–10, 2018.
https://doi.org/10.1007/978-3-319-75025-5_1

practical training. So far, hands-on training has been done primarily on human cadavers, rats or live pigs. Technological advances such as 3D printing, provide novel opportunities for preparation of organ models on which simulation can be performed [11, 17].

VR is a sophisticated computer technology that enables user's physical presence in a virtual or imaginary environment by the generation of realistic images, sounds, and other sensations. This technology has been used in the medical simulation for some time. The widespread implementation of minimally invasive surgical procedures, which have a long and steep learning curve, led to the development of VR simulators for different disciplines and surgical fields like in arthroscopic surgery, endoscopy, intravascular interventions, orthopedics, ophthalmology, spinal surgery, and most recently neurosurgery. Neurosurgery is becoming a new area for simulators, and it seems that it is a good development direction, because of the character of neurosurgical procedures.

2 The Rationale for Simulators in Neurosurgery

Neurosurgery is a medical specialty that deals with the surgical treatment of diseases affecting the central and peripheral nervous system: the brain, spinal cord, peripheral nerves and their cerebrovascular system. Neurological surgery is a highly demanding specialty. In addition to knowledge, experience and the empathy necessary to treat critically ill people – prerequisites in all medical disciplines, neurosurgery requires a particular set of psychophysical skill such as good visual-motor coordination, excellent imagination and spatial awareness. Physical and mental endurance are also essential.

For over hundred years in all surgical specialties, including neurosurgery, a traditional training system has been based on observation. A young adept assists in operations by observing his/hers supervisor's maneuvers until she/he can try to carry them out independently with varying degree of success. Surgical errors can have catastrophic consequences. Learning during surgery also increases the length of medical procedures, increase costs and the overall risk for the patient. This training model has apparent disadvantages: first of all is very time consuming and risky for both parties: the patient and the surgeon. The increased risk must be honestly brought up in the ethical discussion: after all, every patient deserves a competent doctor. A patient has the right to be operated by a fully qualified surgeon and not by a surgeon during training. If this principle were strictly adhered to, it would be practically impossible to train successive generations of surgeons, but also to develop any progress because the problem was not only the training of young adepts but also the introduction of new techniques by experienced and educated doctors. Learning and implementing new methods requires individual learning. There is a limited number of instructors and time limits along the road. The number of cases is also significant. For example, recent advances have made some very technically challenging procedures such as microsurgical aneurysm clipping gradually replaced by less technically demanding endovascular minimally invasive procedures. It can be used in most patients, but not in everyone. So there is a small group that has to be operated classically [12, 16, 19]. It will be increasingly difficult to train neurosurgeons to perform these most technically demanding procedures correctly as the number of "traditional" cases dwindles. It is the reason why neurosurgeons need to practice and improve their skills outside the operating room. Practice in a controlled environment gives a student the opportunity to make mistakes without consequences. However, providing such opportunities for practice poses several challenges. Institutions responsible for medical education in individual countries increasingly recognize the need to implement simulation scenarios as a way to overcome these obstacles. Such simulations will include procedural tasks, crisis management and the introduction of students into clinical situations.

In order to ensure appropriate training opportunities in neurosurgery, plastic, animal and cadavers models are used. Up to now, practical training has taken place mainly on human corpses, human-derived formulations or live rats or pigs [13]. However, they are all less than ideal. Technological advances such as 3D printing, have also provided a wide range of possibilities for the preparation of organ models and their pathologies, where simulations can be made on so-called hybrid models, where in a fixed preparation obtained from human corpses the pathology delivered from 3D printing can be implemented. However, even in this way prepared artificial training model we are not able to completely replace the sensations of natural tissues, and simulation of dynamic situations that take into account even the flow of blood is extremely difficult or impossible.

Thus, plastic models and human anatomical specimens do not have the same properties as living human tissues, and the anatomy of animal models is different from human anatomy. A significant cost of exercise on animal models and cadavers is also the limitation.

It is only recently that an important part of the residential training program in surgical disciplines is the practice of surgical simulators, which are gradually emerging through the cooperation of physicians and engineers.

After all, it is evident and long accepted that in many areas of life in which occupations requiring intense psycho-physical training such as a sports driver or airplane pilot are used for training simulators. There are often very complex systems based on complex computer technologies, sometimes using the most sophisticated information technology. The similar systems have also been utilized for some time in medical education. At many colleges in the world, medical students learn algorithms in medical simulation centers, also based on the so-called augmented virtual reality (AR). VR and AR training simulators also provide a promising opportunity to expand simulation in surgery.

These simulators work analogous to flight simulators, where students spend hours before they are allowed to exercise in the sky. Surgeons also should gain access to tools to practice complicated procedures without exposing patients to unnecessary danger. Also, doctors can practice procedures at any time without the limitations of being available to cases, severity, and location. In addition, VR is a unique source of education with the anatomical structure. One of the greatest challenges in medical education is to provide a realistic sense of interdependence between anatomical structures in 3D. With the VR, learners can re-examine relevant structures, split them, stack and display them from almost any 3D perspective [10].

VR simulation is most often used for training by combining patient data with anatomical information from the atlas to own visualization of known structures. It can be employed for routine training or to focus on particularly difficult cases and new surgical techniques. Possible and most obvious applications include extensions of previously used intraoperative navigation systems such as neuro-endoscopy, stereotaxy, robotic surgery. The final objectives, however, include the simulations of all neurosurgical procedures: neurological and microsurgical procedures. The potential limitations of VR training simulators are related to the transfer of skills from simulation to the real patient. The interaction method should be similar as in the real case to simulate the surgical procedure realistically. Even if this ideal situation is not entirely possible, then you have to strive for it.

But even such imperfect VR system can serve as a valuable part of the anatomical education system. Another important factor in surgical training is the transmission of information between surgeons when evaluating a given data set. Systems have been developed that enable manipulation of 3D data with a large stereoscopic projection system so that the instructor can manipulate the image and share information and insights with a number of recipients in the audience. The role of such systems in teaching surgical strategy and neurosurgical anatomy is invaluable. The person using the virtual reality device can "look around" the artificial world, move and interact with virtual functions or elements. This technology has also been used for some time in simulation surgery. Widespread implementation of minimally invasive surgical procedures, which have a long and steep learning curve, led to the development of simulators VR arthroscopic surgery, endoscopy, the intervention of vascular, orthopedics, ophthalmology, spine surgery, and recently there are attempts to use this technology in neurosurgical education.

The nature of neurosurgical specialty, which requires first and foremost an excellent spatial orientation in neural structures with a high degree of complexity, simulation in virtual reality seems to be the perfect path of development in the mode of training of young neurosurgeons. Initial tests on the simulator manual skills also allow for better pre-selection of candidates for costly training - especially in systems where health training is financed by public funds [1].

Experienced neurosurgeons can also appreciate the benefits of implementing such systems. The ability to create a virtual environment based on actual imaging data offers the opportunity for better preparation and planning of surgery. It may, therefore, be an essential element of the so-called individualized therapy, focusing on specific pathologies in a particular patient.

3 The State of Art in the Field

The use of modern 3-D technology in neurosurgical training and planning has begun with 3-D printed models. These models are primarily used in simulating vascular procedures such as aneurysm clipping. The treatment for cerebral aneurysms to prevent rerupture is performed either with coils (endovascularly) or clips (microsurgically), which are ubiquitously-accepted approaches. After the publication of ISAT study [16], an increasing number of patients receive the endovascular treatment of aneurysms, but still, there is a number of aneurysms for which primary treatment remains microsurgical approach (depending on lesion morphology, their size, and location) [8]. Reduction of the frequency of these microsurgical treatments results in a loss of skills and fewer opportunities for training of residents. Therefore, the demand for developing various efficient and safe surgical training methods has risen under such a circumstance. Several publications have been published in the literature describing the production and testing of three-dimensional cerebral models in this particular medical situation requiring simulation. Mashiko et al. composed models of a trimmed skull, retractable brain, and a hollow elastic aneurysm with its parent artery [15]. The models were created using 3D printers. Residents and junior neurosurgeons attended the training courses. The trainees retracted the brain, observed the parent arteries and aneurysmal neck, selected the proper clip, and clipped the neck of an aneurysm. These trainees succeeded in performing the simulations, and their skills improved in comparison with those exhibited before training.

Wang L et al. compared two types of 3-D aneurysm models: "the regional model" and "the whole model." The regional one included only the aneurysm and adjacent arteries. The whole design included an aneurysm with adjacent arteries, skull base, and nerves. The three-dimensional models were used for surgical planning of the craniotomy and clipping and rated by neurosurgeons via questionnaires. The regional model was judged more realistic for simulating the clipping itself while the "whole 3D model" improved understanding of the surgical view more realistic than the "regional model" did [20].

3D models can also be used in spine surgery as reported by Xu et al. [22]. They used a 3D printer to plan proper insertion of transpedicular screws within the thoracic spine. In their work, they performed models of spine segments in seven post-traumatic patients. Based on the 3D models, surgeons selected the appropriate screw entry points and angles that were then used during the operation. The results suggest that the utilization of the 3D-printed model in preoperative planning improves the outcome.

VR and AR training simulators also provide a promising opportunity to expand simulation in neurosurgery. At present, at least two competing models of neurosurgical simulators have already appeared on the world market. One of the first solutions of this kind in neurosurgery is NeuroTouch® system, which is based on a finite element method and using real-time computing which can assess the multiple features of simulated surgical procedures such as brain tumor resection and healthy tissue associated injury. The elimination of patient risks associated with technical skills learning is the ultimate goal of this simulation-based training [1, 9]. In a safe, simulated environment, the learner achieves the desired learning outcomes where one can repeat the simulated procedure(s) with appropriate demonstrator and performance feedback [4, 18]. The utilization of VR simulators like NeuroTouch® and adequate metric technologies designed to address specific educational, psychomotor, and cognitive issues could improve surgical skill acquisition and assessment, enhance procedural outcomes, and further our understanding of surgical expertise. The NeuroTouch® platform generates output metrics data, which provides quantitative assessment measures used to track and compare psychomotor performance during simulated operative procedures [5].

A critical component of the collaborative studies of the fifteen members of the NeuroTouch® Consortium spread across three continents is the standardization of validated performance metrics [9, 15]. The output data file about a particular task performed on NeuroTouch® contains valuable information concerning psychomotor and cognitive performance. Critical data extraction from this file necessitates the use of sophisticated software and engineering expertise. A pilot validation study of the tumor resection module was conducted in 2014 during which participants expressed the wish for it to be incorporated into their training. The simulator was able to distinguish participants by training level. Being a video game player or musician had no effect on performance, which is important as such associations could skew performance on a simulator such that it may not reflect live surgical performance. NeuroTouch® has been used to assess performance metrics in tumor resection, many of which cannot be measured in live surgery, such as a percentage of tumor removed [4–6]. Furthermore, changing the parameters of the session (for example consistency of tumor, a color of the tumor, and tumor complexity) had a significant influence on the performance of the participants. Interestingly, this change could distinguish between seniority of the performing surgeon, e.g. increasing tumor complexity affected the bimanual psychomotor performance of the residents significantly more than the neurosurgeons [3]. The ability to distinguish participants by training level from these metrics suggests that this simulation environment may be used to evaluate trainees' performance versus their expected performance. There is as yet, little evidence of the usefulness of NeuroTouch® in the acquisition of skills, but these results suggest it may soon be used for learning. The biggest disadvantages of the system are at present limited number of procedures that can be simulated - twelve oncological cases, as well as high price - over hundred thousand USD.

ImmersiveTouch® integrates a haptic device and high-resolution stereoscopic display into VR goggles. It is a joint effort of engineers and neurosurgeons at the University of Illinois at Chicago and the University of Chicago [2]. This simulation platform uses a variety of sensory methods, including visual, auditory, tactile, and kinesthetic, reproducing the sensations experienced during the actual procedure. This is a flexible system that allows you to develop various training applications for different types of skills, not just neurosurgery. Ventriculostomy was the first programmed neurosurgical procedure, but other cranial procedures, such as trepanation, rhizotomy, oncology, spine surgery, such as percutaneous insertion of screws and vertebroplasty, are currently being introduced [7, 14]. The system is still being developed, more affordable from a competitor, but like its predecessor, it does not make it possible to create a virtual environment based on imaging studies of a particular patient.

4 The Description of the Project

The project to be carried out by the interdisciplinary team involves neurosurgeons associated with Vital Medic Education in Kluczbork and Department of Neurosurgery of the University Hospital in Opole is intended to be devoid of most of the drawbacks and disadvantages mentioned above.

The goal of the project is to create a universal neurosurgical simulator (NS) that will be a tool for planning, simulation and training of a specific neurosurgical procedure for a particular patient based on his/her own imaging data such as computer tomography (CT), magnetic resonance imaging (MRI), angiography (DSA) or positron emission tomography (PET). The simulator will enable the import of DICOM volumetric medical imaging data and generation of anatomically accurate patient's skull model rendered by HTC Vive goggles.

Such electronic glasses produce an image with a refresh rate of 90 Hz on two 1080×1200 screens (one for each eye). The device employs more than 70 sensors such as MEMS accelerometers and gyros and laser position sensors, which, in conjunction with lighthouse stations, allow determination of a position of a user with submillimeter accuracy in areas up to 4.6 by 4.6 m. We will use data controllers offered by Manus VR or Neurodigital Technologies. The choice of the controller will be made when the manufacturers start their official distribution (end of 2017). Manus VR glove offers the accuracy of $\pm 3°$. Gloveone gloves from Neurodigital Technologies potentially offer far greater possibilities. In addition to the smart gesture controller, the ten built-in haptic transducers enable you to recognize objects texture and mechanical properties. The user will be able to freely navigate through the intracranial space (translation, rotation, scaling) with the possibility of selecting the anatomical structures to be analyzed, for example, the cerebral vascularization in the localization and supply of vascular malformations. The software will offer 3D annotations, selection and export of region of interest (ROI) for further analysis or 3D printing. The fundamental problem of oncological neurosurgery is the need to remove as much of the pathological focal lesions (for example, tumors) while minimizing potential neurological deficits. Thus, the area of resection should not include the essential nerve fibers of the white matter or the eloquent areas of the cerebral cortex. Diffusion tensor imaging (DTI) - non-invasive magnetic

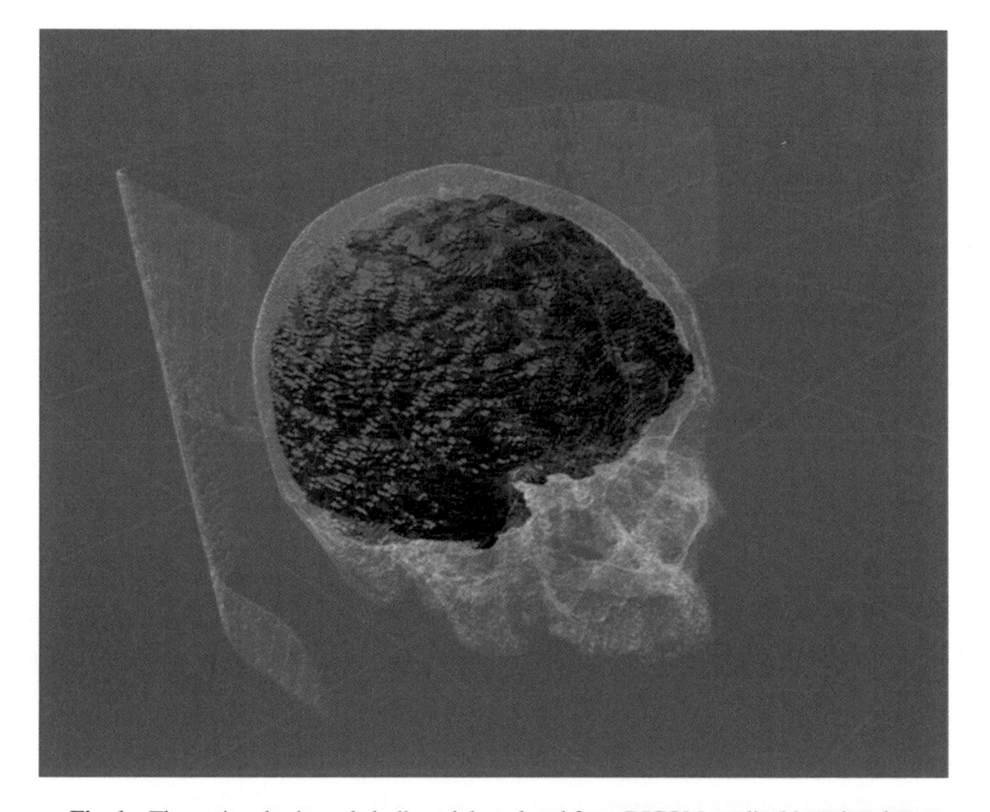

Fig. 1. The patient brain and skull model rendered from DICOM medical imaging data

resonance imaging technique, allows visualization of the direction and continuity of the nerve fibers in vivo. The original and innovative functionality of the neurosurgical simulator will be the visualization of the nerve network from the perspective of the risk of neurological complications. The algorithm developed by the research team will determine optimal access routes to neurosurgical intervention sites that minimize post-operative complications. Using the 3D annotation system, the user will be able to mark the volume of resection and visualize it against the background of the surrounding blood vessels and nerve fibers. The VR images will be generated from the perspective of the operator taking into account the position of the patient on the operating table and magnification of microscopes used in clinical practice (Fig. 1).

5 Conclusion

The younger generation of neurosurgeons need VR neurosurgical simulators to facilitate their educational development. Experienced neurosurgeons can employ them for surgery planning, especially in the fields of rare pathologies. These needs is the rationale for close cooperation between neurosurgeons and biomedical engineers.

References

1. Alaraj, A., Charbel, F.T., Birk, D., Tobin, M., Luciano, C., Banerjee, P.P., Rizzi, S., Sorenson, J., Foley, K., Slavin, K., Roitberg, B.: Role of cranial and spinal virtual and augmented reality simulation using immersive touch modules in neurosurgical training. Neurosurgery **72**(1), 115–123 (2013). https://doi.org/10.1227/neu.0b013e3182753093
2. Alaraj, A., Luciano, C.J., Bailey, D.P., Elsenousi, A., Roitberg, B.Z., Bernardo, A., Banerjee, P.P.: Charbel FT: virtual reality cerebral aneurysm clipping simulation with real-time haptic feedback. Neurosurgery **2**, 52–58 (2015). https://doi.org/10.1227/neu.0000000000000583
3. Alotaibi, F.E., AlZhrani, G.A., Mullah, M.A., Sabbagh, A.J., Azarnoush, H., Winkler-Schwartz, A., Del Maestro, R.F.: Assessing bimanual performance in brain tumor resection with NeuroTouch, a virtual reality simulator. Neurosurgery **11**(Suppl 2), 89–98 (2015). https://doi.org/10.1227/neu.0000000000000631. discussion 98
4. Alotaibi, F.E., AlZhrani, G.A., Sabbagh, A.J., Azarnoush, H., Winkler-Schwartz, A., Del Maestro, R.F.: Neurosurgical assessment of metrics including judgment and dexterity using the virtual reality simulator NeuroTouch (NAJD metrics). Surg. Innov. **22**(6), 636–642 (2015). https://doi.org/10.1177/1553350615579729
5. AlZhrani, G., Alotaibi, F., Azarnoush, H., Winkler-Schwartz, A., Sabbagh, A., Bajunaid, K., Lajoie, S.P., Del Maestro, R.F.: Proficiency performance benchmarks for removal of simulated brain tumors using a virtual reality simulator NeuroTouch. J Surg. Educ. **72**(4), 685–696 (2015). https://doi.org/10.1016/j.jsurg.2014.12.014
6. Azarnoush, H., Alzhrani, G., Winkler-Schwartz, A., Alotaibi, F., Gelinas-Phaneuf, N., Pazos, V., Choudhury, N., Fares, J., DiRaddom, R., Del Maestro, R.F.: Neurosurgical virtual reality simulation metrics to assess psychomotor skills during brain tumor resection. Int. J. Comput. Assist. Radiol. Surg. **10**(5), 603–618 (2015). https://doi.org/10.1007/s11548-014-1091-z
7. Banerjee, P.P., Luciano, C.J., Lemole Jr., G.M., Charbel, F.T., Oh, M.Y.: Accuracy of ventriculostomy catheter placement using a head- and hand-tracked high-resolution virtual reality simulator with haptic feedback. J. Neurosurg. **107**(3), 515–521 (2007)

8. Bekelis, K., Gottlieb, D.J., Su, Y., O'Malley, A.J., Labropoulos, N., Goodney, P., Lawton, M.T., MacKenzie, T.A.: Comparison of clipping and coiling in elderly patients with unruptured cerebral aneurysms. J. Neurosurg. **126**(3), 811–818 (2017). https://doi.org/10.3171/2016.1.JNS152028
9. Clark, A.D., Barone, D.G., Candy, N., Guilfoyle, M., Budohoski, K., Hofmann, R., Santarius, T., Kirollos, R., Trivedi, R.A.: The effect of 3-dimensional simulation on neurosurgical skill acquisition and surgical performance: a review of the literature. J. Surg. Educ. **22**(16) (2017). https://doi.org/10.1016/j.jsurg.2017.02.007. S1931–7204, 30316-6 (Epub ahead of print)
10. Gasco, J., Holbrook, T.J., Patel, A., Smith, A., Paulson, D., Muns, A., Desai, S., Moisi, M., Kuo, Y.F., Macdonald, B., Ortega-Barnett, J., Patterson, J.T.: Neurosurgery simulation in residency training: feasibility, cost, and educational benefit. Neurosurgery **73**(Suppl 1), 39–45 (2013). https://doi.org/10.1227/NEU.0000000000000102
11. Konakondla, S., Fong, R., Schirmer, C.M.: Simulation training in neurosurgery: advances in education and practice. Adv. Med. Educ. Pract. **14**(8), 465–473 (2017). https://doi.org/10.2147/AMEP.S113565
12. Korja, M., Kivisaari, R., RezaiJahromi, B., Lehto, H.: Size and location of ruptured intracranial aneurysms: consecutive series of 1993 hospital-admitted patients. J. Neurosurg. **2**, 1–6 (2016). https://doi.org/10.3171/2016.9.JNS161085
13. Krähenbühl, S.M., Čvančara, P., Stieglitz, T., Bonvin, R., Michetti, M., Flahaut, M., Durand, S., Deghayli, L., Applegate, L.A., Raffoul, W.: Return of the cadaver: key role of anatomic dissection for plastic surgery resident training. Medicine (Baltimore) **96**(29), e7528 (2017). https://doi.org/10.1097/MD.0000000000007528
14. Luciano, C.J., Banerjee, P.P., Bellotte, B., Oh, G.M., Lemole Jr., M., Charbel, F.T., Roitberg, B.: Learning retention of thoracic pedicle screw placement using a high-resolution augmented reality simulator with haptic feedback. Neurosurgery **69**(Suppl Operative 1), 14–19 (2011). https://doi.org/10.1227/neu.0b013e31821954ed
15. Mashiko, T., Kaneko, N., Konno, T., Otani, K., Nagayama, R., Watanabe, E.: Training in cerebral aneurysm clipping using self-made 3-dimensional models. J. Surg. Educ. **74**(4), 681–689 (2017). https://doi.org/10.1016/j.jsurg.2016.12.010
16. Molyneux, A.J., Kerr, R.S., Yu, L.M., Clarke, M., Sneade, M., Yarnold, J.A., Sandercock, P.: International subarachnoid aneurysm trial (ISAT) of neurosurgical clipping versus endovascular coiling in 2143 patients with ruptured intracranial aneurysms: a randomised comparison of effects on survival, dependency, seizures, rebleeding, subgroups, and aneurysm occlusion. Lancet **366**(9488), 809–817 (2005)
17. Mundschenk, M.B., Odom, E.B., Ghosh, T.D., Kleiber, G.M., Yee, A., Patel, K.B., Mackinnon, S.E., Tenenbaum, M.M., Buck II, D.W.: Are residents prepared for surgical cases? Implications in patient safety and education. J. Surg. Educ. **18** (2017). https://doi.org/10.1016/j.jsurg.2017.07.001. S1931-7204
18. Pfandler, M., Lazarovici, M., Stefan, P., Wucherer, P., Weigl, M.: Virtual reality based simulators for spine surgery: a systematic review. Spine J. **17** (2017). https://doi.org/10.1016/j.spinee.2017.05.016. S1529-9430, 30208-5. (Epub ahead of print)
19. Steklacova, A., Bradac, O., Charvat, F., De Lacy, P., Benes, V.: "Clip first" policy in the management of intracranial MCA aneurysms: Single-centre experience with a systematic review of the literature. Acta Neurochir. (Wien) **158**(3), 533–546 (2016). https://doi.org/10.1007/s00701-015-2687-y. discussion 546
20. Wang, L., Ye, X., Hao, Q., Chen, Y., Chen, X., Wang, H., Wang, R., Zhao, Y., Zhao, J.: Comparison of two three-dimensional printed models of complex intracranial aneurysms for surgical simulation. World Neurosurg. **103**, 671–679 (2017). https://doi.org/10.1016/j.wneu.2017.04.098

21. Winkler-Schwartz, A., Bajunaid, K., Mullah, M.A., Marwa, I., Alotaibi, F.E., Fares, J., Baggiani, M., Azarnoush, H., Zharni, G.A., Christie, S., Sabbagh, A.J., Werthner, P., Del Maestro, R.F.: Bimanual psychomotor performance in neurosurgical resident applicants assessed using neurotouch, a virtual reality simulator. J. Surg. Educ. **73**(6), 942–953 (2016). https://doi.org/10.1016/j.jsurg.2016.04.013
22. Xu, W., Zhang, X., Ke, T., Cai, H., Gao, X.: 3D printing-assisted preoperative plan of pedicle screw placement for middle-upper thoracic trauma: a cohort study. BMC Musculoskelet Disord. **18**(1), 348 (2017). https://doi.org/10.1186/s12891-017-1703-1

Outlier Correction in ECG-Based Human Identification

Volodymyr Khoma[1,2(✉)], Mariusz Pelc[1], Yuriy Khoma[2], and Dmytro Sabodashko[2]

[1] Opole University of Technology, 76 Proszkowska Street, 45-758 Opole, Poland
{v.khoma,m.pelc}@po.opole.pl
[2] Lviv Polytechnic National University, 12 Bandera Street, Lviv 79013, Ukraine
khoma.yuriy@gmail.com
http://www.po.opole.pl/en/
http://www.lp.edu.ua/en/lp

Abstract. In this article we have proposed a novel method for ECG signal processing in biometric applications. The main idea is to correct anomalies in various segments of ECG waveform rather than skipping a corrupted ECG heartbeat, as it is commonly done in most cases. The proposed approach is taking into consideration that biosignals are of quasi-periodic nature.

Neighbouring ECG heartbeats are analysed using a sliding window. Within such a window the analysis of samples distributions is being performed. This information allows to detect outlying samples and correct them with expected values. Such an approach allows to collect better statistical representation which improves identification models performance.

In order to validate our method we used open-source Physionet ECG-ID database. This database contains 310 records per 90 unique persons. The classification result reported on this data set using commonly known outlier detection approach was 91%. We carried out a number of experiments and then compared the obtained results to those obtained using the outlier correction method described above. Classification results for our method exceeded 95%. Thus, misclassification error rate has been improved twice.

Keywords: Outlier correction · Anomaly detection · Biometrics
Human identification · ECG · Physionet · ECG-ID database

1 Introduction

Electrocardiogram (ECG) is representation of bio-electrical heart activity. ECG signal is typically registered using electrodes as voltage difference appearing on body surface. Because of very low voltage levels, the ECG signal is sensitive to various kinds of disturbances (artefacts related to muscles activity, power supply related disturbances and others) [1]. In order to guarantee appropriate quality of the registered ECG signal, in the clinical practice the following conditions are typically satisfied: overall body-to-electrode conductivity is being improved by the use of a conductive gel, multichannel bio-potentials acquisition from different parts of chest and limbs (typically 12 leads) is used, comfortable conditions throughout the acquisition process (patient is laying on a sofa), also, patients are typically holding their breath for the registering duration [2, 3].

W. P. Hunek and S. Paszkiel (Eds.): BCI 2018, AISC 720, pp. 11–22, 2018.
https://doi.org/10.1007/978-3-319-75025-5_2

Recently the ECG signal is not only used for medical purposes (like e.g. widely used diagnostics) but also for other purposes, e.g. in biometric for persons identification. In [4, 5] the authors discuss a number of ECG signal features that carry individual properties of a human being:

- originating from a vital and living organ,
- more reliable and robust due to their internal and not external biometric making it very difficult to mimic or forge,
- allowing to provide a fresh biometric readings continuously.

The aims of biometrics differ from aims of medical diagnostics. The conditions of ECG signal acquisition are also different. In biometrics the main focus is on not very sophisticated signal acquisition from a person, for example from the first lead using dry electrodes measuring signals from fingers on the right and left hand. There is also a possibility to acquire the ECG signals from chest using T-shirt with embedded textile electronic(s) or from neck using necklace and a necklet [6, 7]. Similar non-intrusive methods of the ECG signal recording are becoming more and more useful for regular application and also widens the range of their potential application.

A commonly known problem is that typically the dry electrodes do not guarantee appropriate quality of the acquired ECG signal. In order to increase the signal/noise ratio (SNR) various kinds of digital signal processing methods (DSP) are used and most of all – various filtration methods. Conventional filtration methods ensure only efficient attenuation of harmonics (50 or 60 Hz – notch filters) sourced from power supply network or noises from outside of useful signal.

However, the filtration process in itself results in some remaining distortions and to eliminate them some very advanced signal processing and analysis methods are being used. These methods first detect deviations from norm and further elimination of anomalous segments from the ECG recording. As a result this increases the level of correct decisions made by system of biometric identification.

Until today there have been proposed numerous outlier detection methods and algorithms. In machine learning terminology these methods can be divided into to classes [8, 9]:

- supervised, meaning that these methods are based on the "normal" data model and observing the level of deviation of the analysed ECG signal recording from that model,
- unsupervised, meaning that these methods do not require prior knowledge of data characteristic, actually, their operation is based on finding condensed data groups and referring the analysed recording to one of such groups (clustering-based classification).

All known outlier detection methods are used for eliminating those heartbeats that have been qualified as anomalous [10]. Typically, in the biometrics applications the ECG recordings are not very long (from a few up to several dozens of heartbeats) and hence removing parts of abnormal heartbeats leads to reducing the data set being subject to classification. As a result it may have a negative impact on the performance of the biometrical identification system.

Authors of this paper propose a completely new approach toward improved ECG-based human identification system and rather than removing the abnormal ECG signal beats, we propose to correct even those records that were assessed as abnormal even after pre-processing stage.

The aim of this paper is to present the Authors' method for detection and correction of anomalies in the ECG recordings as well as its validation and comparison to existing methods.

The remaining part of the paper is structured as follows: in Sect. 2 the structure of human identification system based on the ECG signals is described along with description of essence of signal processing at different stages. Section 3 presents the proposed method of the ECG signal segments outlier detection and correction together with its implementation details. Sections 4 and 5 includes description of the methodology and results of outlier detection and correction methods in application to real ECG signals followed by some concluding statements.

2 Human Identification System Based on ECG

The typical process of human identification based on ECG analysis consisting of the following phases: ECG signal acquisition, signal and data pre-processing, feature extraction, feature reduction and classification [10–12]. The block diagram of such a human identification system based on ECG is shown in Fig. 1.

Systems of biometrical identification commonly known from literature use various algorithms of data processing at respective stages [12–14]. Because the main aim of this paper primarily is to estimate the efficiency of the proposed outlier detection and correction method, for the comparison purposes the Authors have chosen the research results described in [12]. This choice was resulting from the two following factors: ECG data base availability [15] as well as a very detailed description of the research methodology making possible the verification of the presented results.

Furthermore, while our research was carried out, we have come to the conclusion that the algorithms used at the pre-processing stage in [12] can be simplified. Main features of the algorithms have been outlined below.

ECG acquisition. Due to the practical reasons, the input signal to the identification system was measured as difference between bio-potentials between the left and right hand (so called I-lead). The data acquired during this research included in the ECG-ID Database comprise of 310 recordings of ECG signals from 90 persons [15]. Each recording is of 20 s long and was sampled with the frequency of 500 Hz at 12-bit precision.

Signal and data pre-processing. The aim of this stage is improvement of the quality of the raw ECG signal, usually being subject of disturbances of different nature. In [12] the baseline drift correction was being done using wavelet transform (wname = 'db8', N = 9). Adaptive band stop (notch) filter (Ws = 50, dA = 1.5) was used to attenuate the power-line noise. Additionally, remaining noises were eliminated from the raw ECG signal using digital low-pass Butterworth filter with the following parameters:

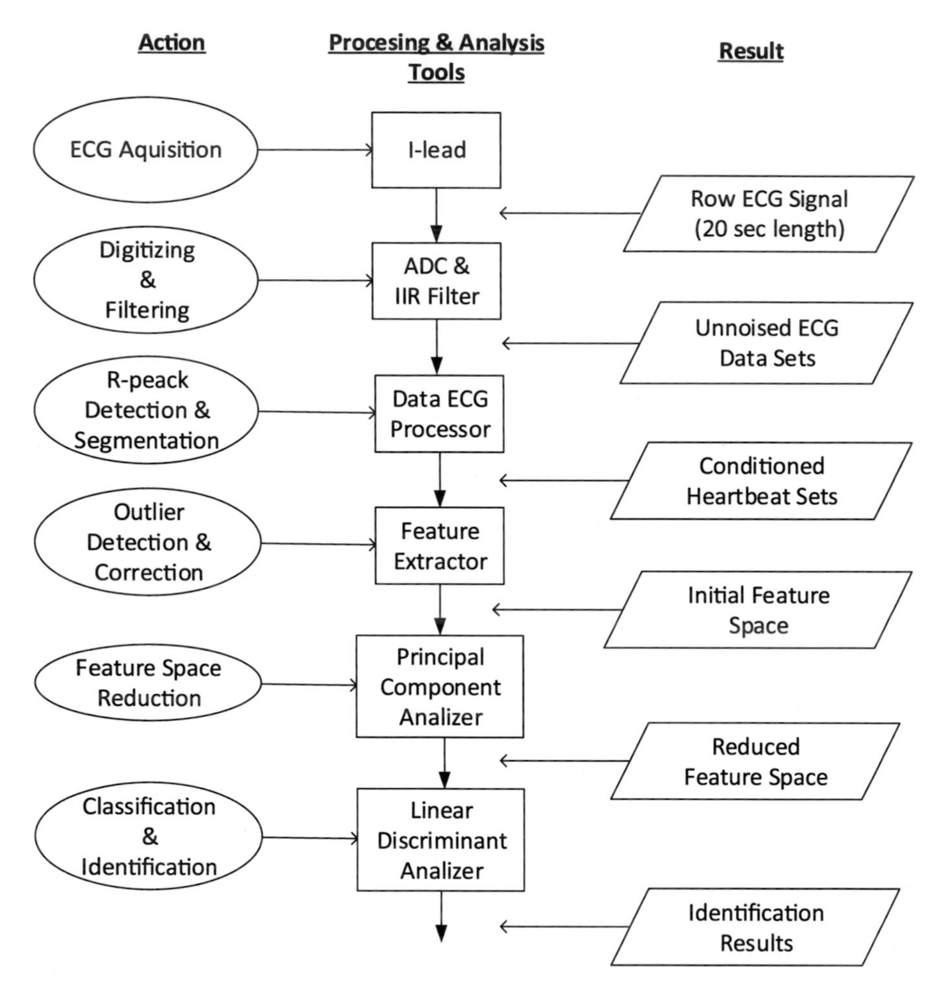

Fig. 1. The block diagram of human identification system based on ECG

Wp = 40 Hz, Ws = 60 Hz, Rp = 0.1 dB, Rs = 30 dB. In this work we present the possibility of elimination of deformations using IIR filter with parameters further described in Sect. 4. The proposed pre-processing solution is computationally less complex yet guaranteeing comparable identification results.

The cardiac electrical activity information is contained in the cardiac cycle segment containing the central QRS complex, opening P and closing T waves [1]. For the segmentation purpose usually there are used various R-peaks detection algorithms, for example based on Hamilton algorithm or Pan-Tompkins [16]. Furthermore, changeability of the analysed segments can be reduced based on normalisation of the ECG signal.

Feature extraction. In order to form the initial feature space from the ECG segments, Authors have proposed their own solution relying on detection of anomalous sections in ECG segments and their corrections. Details of the proposed solution are further

explained in Sect. 3. As a result application of this method the initial feature space contains 250 samples in 12-bit format.
Feature reduction and classification. In order to decrease the computational complexity of the classifier, feature reduction operation was performed using well known algorithm – Principal Component Analysis [17]. The classifier was also built based on well known Linear Discriminant Analysis algorithm [18]. Both algorithms have proven their effectiveness in the research carried out in [12].

3 Outlier Correction

In order to improve the classifier operation, Authors propose the presented below method for detection and correction of anomalies in the ECG segments. Those anomalies are usually arising as a result of inadequacies of the measurement process. Example anomalies are shown in Fig. 2.

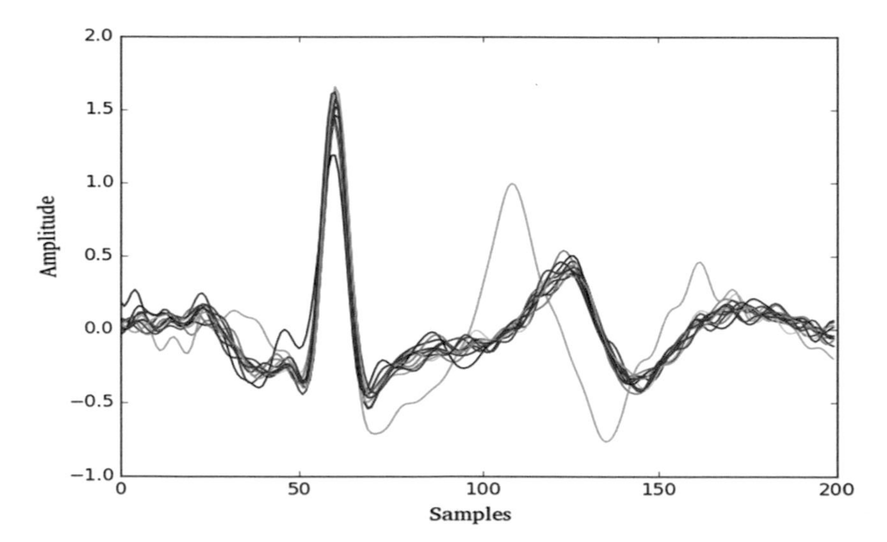

Fig. 2. Segmented ECG signal including of sizable (blue curve) and small size (black curve) anomalies

In practice the ECG segments (heartbeats) that contain big anomalies can be relatively easy to detect. In most cases for this purpose the Euclid distance estimation is used and typically an erroneous segment is being dropped even if the anomaly affects small amount of samples. It is worth emphasising that even for small anomalies their detection is extremely important because they can disturb the classifier operation. The advantage of the proposed method is that it remains valid for detection of both, small and big anomalies. Furthermore, in the contrary to some other known solutions, the proposed method allows to correct the anomalies without the need of eliminating the whole heartbeat. This is of utmost importance from the point of view of keeping the sufficient number of heart beats necessary for the biometric system to work correctly.

The main idea of the proposed method is as follows:

1. Finding anomalous sections (windows) in each heartbeat, within which deviation of even one sample exceeds a defined threshold.
2. Replacing those anomalous sections with equivalent sections acquired as a result of averaging of all heartbeats.

In the first stage, namely outlier detection, it is being defined a vector of average values for each sample

$$\bar{x}(n) = \frac{1}{K}\sum\nolimits_{k=1}^{K} x(k,n), \tag{1}$$

Where $x(k,n)$ – element of the ECG record data matrix $\mathbf{X}(K,N)$; $k \in 1 \div K$ – rows representing number of heartbeats; $n \in 1 \div N$ – columns representing number of samples per beat.

Also, the vector of standard deviations is calculated using formula (2)

$$\mathrm{std}(n) = \frac{1}{K-1}\sqrt{\sum\nolimits_{k=1}^{K} \mathrm{x}(k,n) - \bar{\mathrm{x}}(n)}. \tag{2}$$

Further, average value of standard deviation for matrix $\mathbf{X}(K, N)$ can be calculated as

$$\overline{\mathrm{std}} = \frac{1}{N}\sum\nolimits_{n=1}^{N} \mathrm{std}(n). \tag{3}$$

At last, the original matrix $\mathbf{X}(K,N)$ of samples of the ECG record can be replaced with a binary matrix $\mathbf{O}(K, N)$ the same dimension

$$\mathrm{o}(k,n) = \frac{\mathrm{abs}[\mathrm{x}(k,n) - \bar{\mathrm{x}}(n)]}{\overline{\mathrm{std}}} > gain, \tag{4}$$

where $o(k,n)$ – element of the binary matrix $\mathbf{O}(K,N)$, where each non-zero value represents the detected outlier (for a given index of sample and record).

In the second stage, namely outlier correction, following transformation is performed:

$$x(k, n{:}n+L) = \begin{cases} \bar{\mathrm{x}}(n{:}n+L), \text{if any in } o(k, n{:}n+L) = 1 \\ x(k, n{:}n+L), \text{if all in } o(k, n{:}n+L) = 0 \end{cases}, \tag{5}$$

where $\bar{\mathrm{x}}(n{:}n+L)$ – all averaged samples within window of length L,
$x(k, n{:}n+L)$ – all samples of record k, within window of length L,
$o(k, n{:}n+L)$ – all outliers indexes of record within window of length L.
The pseudocode of the algorithm fulfilling the proposed method is shown in Fig. 3.

```
function  outliers_correction (ecg_records)
Input: The ECG record data matrix with dimension of KxN
(number of heartbeats * number of samples per beat)
Output: The outlier corrected matrix with the same dimen-
sion as the input one

mean_vector ← mean values for each sample (eq. 1)
std_vector← standard deviations for each sample (eq. 2)
averaged_std  ← mean value of std_vector (eq. 3)

# outlier detection stage (eq. 4)
for beat inecg_records
outliers[beat, :] ←abs(ecg_records[beat, :] -
mean_vector)/averaged_std> gain
end

# outlier correction stage (eq. 5)
for beat in ecg_record
for window in beat
ifany(outliers[beat, window] istrue)
ecg_records[beat,window] ← mean_vector [window]
  end
end
end
```

Fig. 3. Windows outliers correction algorithm

The described algorithm is using two hyper parameters: *gain* and *window length*. As a result, although the hyper parameters are identical across the whole base, the outlier criteria is being adaptively changed and adjusted for each record, as it takes into consideration averaged standard deviation value.

The choice of parameters is being done in the following manner:

1. Rough tuning is being performed visually in application to a part of the data, which allows to detect and correct some big anomalies. The main results of this stage are hyper parameters values that are then used as initial point for further optimization (fine tuning),
2. Fine tuning is being performed along with the rough tuning. The main goal of this stage is to find such values of the hyper parameters that would lead to minimal classification error. In fact, this is a conventional numerical optimization task.

It is worth emphasising that the algorithm accuracy is strongly dependant on how well the initial values of the hyper parameters were selected. In case the gain is too much loose-fitting, the whole correction algorithm may even entirely loose its accuracy, leaving deformed segments in data sets fed into the classifier. On the other hand, in case the gain is too much stiff-fitting, too many segments of the actual data may be replaced with the segment averaged within a record. So, any improper choice of parameters results leads to improper operation of the whole algorithm of anomalies correction.

In Fig. 4. one can see results of the operation of the proposed correction algorithm using the ECG signal shown in Fig. 2.

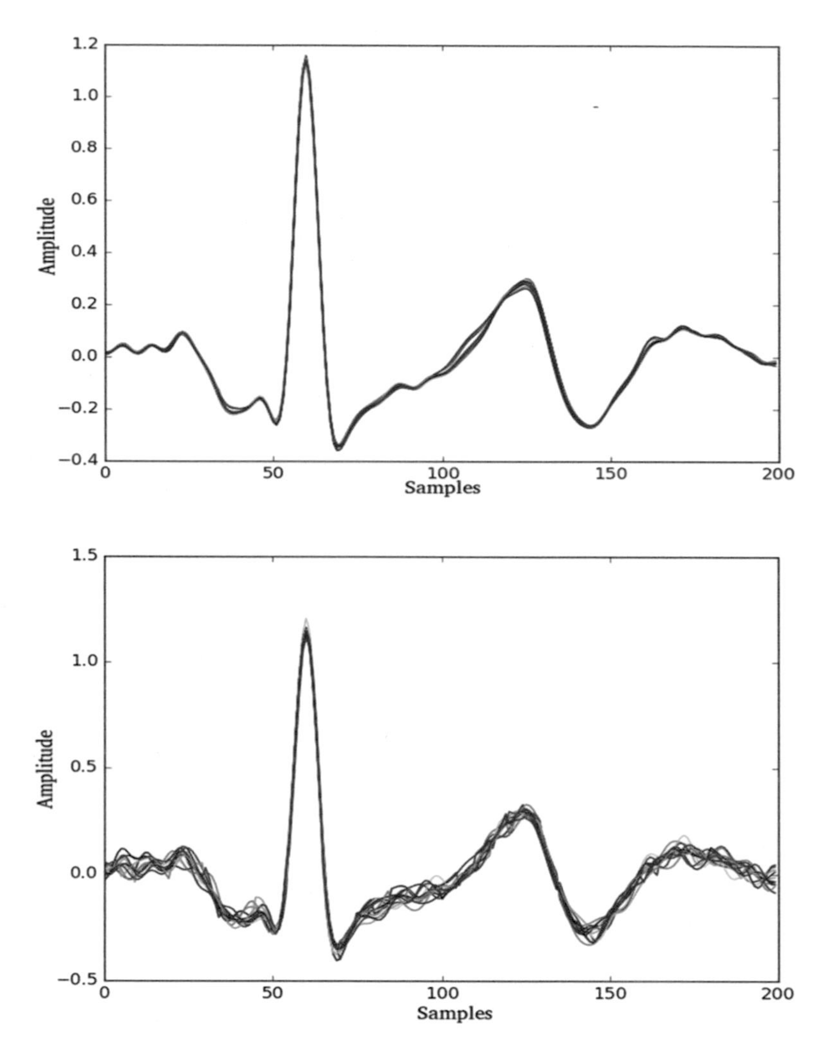

Fig. 4. Results of the operation of the anomaly detection and correction algorithm with stiff-fitting gain (above) and visual assessment based gain (below)

According to our research it becomes evident that the sections of the ECG signal that are subject of the most frequent corrections are the P and T waves. The number of corrections though varies depending on the class of the records. On the records of one class there can be by average 2–3 corrections per 10 heartbeats whilst on some other class records it can even be 18–20 corrections. Each of the corrected segments can be used as a separate template for the identification system classifier.

4 Experimental Results

In order to estimate the efficiency of outliers correction in biometrics applications it is reasonable to compare the proposed method with other methods, for example the one described in study [12]. The study was conducted on the open database and it provides a very detailed description of the experiment. These are two main reasons to choose it as baseline for further comparison and analysis.

At first, it was planned to reproduce the exact experimental results as described in the paper. However, it was implemented with minor modifications because of following reasons:

1. Training and test data. This publication does not describe all the details of how records were split into training and test subsets. General idea is that "differentiation between the training and test sets aimed to provide maximum performance complexity, i.e., maximum difference between records in different sets both in monitoring time and human physical state" [12]. This statement is unfortunately not very precise and as such it does not allow to reproduce the results one to one. So, in our experiments, in order to achieve the fair performance estimation all data have been split randomly for 30 times. The distribution with the lowest accuracy was then being chosen for further investigations as the worst case scenario.
2. Heart rate correction. The paper indicated that heart rate may very strongly for different records. To address to this issue the raw ECG waveform was modified in a way to obtain a heart rate of 60 beats per second for each record. For this purpose the Framingham or Bazett's formulas were considered. Although these formulas are well known in cardiology they are typically used to normalize the duration of QT interval[1] as a parameter, which can indicate some health problems. Thus from a medical point of view there is no need to adjust the waveform of ECG signal itself. So this stage was entirely omitted and records without heart rate correction were used for further experiments.
3. Signal pre-processing. In [12] the complex and multilevel process of drifts correction is described together with harmonic attenuation and noise reduction. The paper does not specify the criteria that were used to select the particular pre-processing algorithms and more importantly, how this choice affected the overall accuracy of the classification/identification. Therefore, pre-processing described in [12] was replaced with simple IIR bandpass filter. The filter parameters were selected as follows: polynomial type – Butterworth, sampling rate - 500 Hz, stopband - below 1 Hz and above 50 Hz, passband - between 4 and 35 Hz, stopband gain - 20 dB, passband gain - 1 dB. The filter was designed using SciPy library [19]. Heartbeat segmentations were performed using Hamilton algorithm from BioSPPy library [20].

[1] QT interval is defined from the beginning of the QRS complex to the end of the T wave and represents total ventricular activity (depolarization and repolarization).

Experiments have proven that the classification accuracy remains the same in both cases, but the IIR filter has much lower computational complexity. Therefore, it is reasonable to use this approach for ECG signals pre-processing.

Linear Discriminant Analysis (LDA) was used as classification algorithm. Dimension of the input data was reduced from 270 to 30 features (samples) using PCA transformation (Principal Component Analysis). Experiments were considered in the following circumstances:

1. No outlier processing,
2. Outlier detection described at [12],
3. Outlier correction algorithm describe above (for different hyperparameters).

All experiments have been carried out using Python 2.7. Furthermore, the following frameworks and libraries have been used: SciPy, NumPy, matplotlib, sci-kit learn. The source code of the project can be found here [21]. To estimate the classification performance the accuracy score was used [22]. Results are gathered in the Table 1.

Table 1. Classification results

Outlier processing algorithm	No outlier processing	Processing described in [12]	Proposed approach				
			length = 5 samples			gain = 0.5	
			gain = 5	gain = 0.5	gain = 0.05	length = 3	length = 10
Train set accuracy, %	93.07	99.17	97.66	**99.33**	99.19	99.07	99.15
Test set accuracy, %	84.37	91.22	92.80	**95.18**	93.77	95.03	93.88

Performance of the classification systems is presented for both, train and test sets. It's quite common for most of the machine learning applications. Train set demonstrates a theoretical limit for accuracy that can be achieved by chosen algorithm on a dedicated data set. Test set gives more realistic performance estimation in real-life applications, which can be expected when classifier will operate on a new and previously unseen data. Grid search algorithm has been chosen for hyper parameter optimization. Results gathered in the Table 1 show how sensitive the developed algorithm is to each parameter change. Values presented in the Table 1 cover the range around the found optimum, which is: length = 5, gain = 0.5. Error rate was reduced for more than twice from 9% to 4% in comparison to the approach proposed in [12].

5 Conclusions

Biometrics is the science that uses statistical methods to recognize the human identity based on the physiological and behavioural attributes of the individual such as fingerprints, face and voice recognition, etc. However, huge ubiquity of Internet of Things and wearable solutions enables to extend existing system with new channels. One of the most promising examples is electrocardiogram signal. Plenty of studies have been done in this area. Typically ECG based biometric systems consists of three major stages: data preprocessing, feature extraction/selection and classification.

Data pre-processing stage is focused on drift correction, de-noising, harmonic distortions attenuation and outlier detection and thus is very critical to develop some reliable and robust identification systems. The aim of this paper was to develop and validate a new algorithm for detection and further correction of abnormal ECG heartbeats (outliers).

According to this algorithm, 10 neighbouring heartbeats are analysed simultaneously using sliding window (without overlap). Then samples distribution is analysed within each window. If samples deviation for a separate heartbeat exceeds the threshold they are considered as outliers. After outliers have been detected they are replaced with expected values from neighbouring heartbeats.

To estimate the efficiency of outliers correction in biometrics applications we found it reasonable to make comparison with the study described at [12]. The study was conducted on the open ECG-ID database and experiment description is available in details which were the two main reasons to choose it as baseline for further analysis.

Experiment presented in [12] was reproduced, although with some minor alterations. One of the most important observations was that rather than using separate algorithms for de-noising, drift correction, etc. a simple bandpass IIR filter can be applied without significant impact on the resulting performance.

Experiments have proven that outlier leads to approximately 15% error rate. For the approach proposed in [12] the error rate is less than 9%. The outlier correction method proposed in this paper allows to reduce the classification error down to 5%. Experiments were performed for different hyper parameters. Furthermore, it turned out that the threshold value is more critical parameter than the window size. Empirically, the best configuration is window length of 5 samples and threshold gain 0.5.

The following ideas can potentially be helpful to improve the proposed outlier correction approach:

- use median instead of mean value to estimate samples deviation,
- add recursive loop to outlier correction process (run correction multiple times, should a need arise),
- create some sort of analytical framework for hyper parameters selection.

References

1. Jenkins, D., Gerred, S.: ECGs by Example, 3rd edn., 238 p. Elsilver (2011)
2. Aslanger, E., Yalin, K.: Electromechanical association: a subtle electrocardiogram artifact. J. Electrocardiol. **45**(1), 15–17 (2012)
3. Urigüen, J.A., Garcia-Zapirain, B.: EEG artifact removal-state-of-the-art and guidelines. J. Neural Eng. **12**(3), 031001 (2015)
4. Fratini, A., Sansone, M., Bifulco, P., Cesarel, M.: Individual identification via electrocardiogram analysis. BioMed. Eng. OnLine **14**, 1–23 (2015)
5. Kaur, G., Singh, D., Kaur, S.: Electrocardiogram (ECG) as a biometric characteristic: a review. Int. J. Emerg. Res. Manage. Technol. **4**(5), 202–206 (2015)
6. Shen, T.W., Tompkins, W.J., Hu, Y.H.: One-lead ECG for identity verification. In: Proceedings of the 2nd Joint Conference on the IEEE Engineering in Medicine and Biology Society and the Biomedical Engineering Society, vol. 1, pp. 62–63 (2002)

7. Cunha, J., Cunha, B., Xavier, W., Ferreira, N., Pereira, A.: Vital-Jacket: a wearable wireless vital signs monitor for patients' mobility. In: Proceedings of the Avantex Symposium (2007)
8. Tax, D., Duin, R.: Outliers and data descriptions. In: Proceedings 7th Annual Conference on the Advanced School for Computing and Imaging (ASCI) (2001)
9. Hodge, V.J., Austin, J.: A survey of outlier detection methodologies. Artif. Intell. Rev. **22**, 85–126 (2004)
10. Lourenço, A., Silva, H., Carreiras, C., Fred, A.: Outlier detection in non-intrusive ECG biometric system. In: Kamel, M., Campilho, A. (eds.) ICIAR 2013. LNCS, vol. 7950, pp. 43–52. Springer, Heidelberg (2013). https://doi.org/10.1007/978-3-642-39094-4_6
11. Matos A.C., Lourenc A., Nascimento, J.: Embedded system for individual recognition based on ECG Biometrics. In: Proceedings of the Conference on Electronics, Telecommunications and Computers – CETC, pp. 265–272 (2013)
12. Lugovaya, T.S.: Biometric human identification based on electrocardiogram. [Master's thesis] Faculty of Computing Technologies and Informatics, Electrotechnical University "LETI", Saint-Petersburg, Russian Federation, June 2005
13. AlMahamdy, M., Bryan Riley, H.: Performance study of different denoising methods for ECG signals. In: Proceedings of the 4th International Conference on Current and Future Trends of Information and Communication Technologies in Healthcare (ICTH-2014), pp. 325–332 (2014)
14. Varshney, M., Chandrakar, C., Sharma, M.: A survey on feature extraction and classification of ECG signal. Int. J. Adv. Res. Electr. Electron. Instrum. Eng. **3**(1), 6572–6576 (2014)
15. https://physionet.org/physiobank/database/ecgiddb. Accessed 10 Apr 2017
16. Pan, J., Tompkins, W.J.: A real-time QRS detection algorithm. IEEE Trans. Biomed. Eng. **32**(3), 230–236 (1985)
17. Duda, R.O., Hart, P.E., Stork, D.H.: Pattern Classification, 2nd edn. Wiley Interscience (2000)
18. Jolliffe, I.T.: Principal Component Analysis. Springer Series in Statistics, 2nd edn., 487 p. Springer, New York (2002)
19. Python-based ecosystem of open-source software. https://www.scipy.org. Accessed 12 Apr 2017
20. BioSPPy - Biosignal Processing in Python. A toolbox for biosignal processing written in Python. https://github.com/PIA-Group/BioSPPy. Accessed 12 Apr 2017
21. ecg-identification: Package to process common ECG signals for human identification purposes. https://github.com/YuriyKhoma/ecg-identification. Acceseed 14 Apr 2017
22. Scikit-Learn Machine Learning in Python. http://scikit-learn.org/stable/modules/model_evaluation.html#accuracy-score. Accessed 21 Apr 2017

Application of sEMG and Posturography as Tools in the Analysis of Biosignals of Aging Process of Subjects in the Post-production Age

Zbigniew Borysiuk(✉), Mariusz Konieczny, Krzysztof Kręcisz, and Paweł Pakosz

Faculty of Physical Education and Physiotherapy, Opole University of Technology, Opole, Poland
{z.borysiuk,m.konieczny, k.krecisz,p.pakosz}@po.opole.pl

Abstract. Surface electromyography (sEMG) and posturography are some of the most useful tools applied in the assessment of the motor skills of humans. The scope and objective of this paper is to report on concurrent and synchronized application of the Kistler platform for postural balance and electromyography analysis (sEMG Noraxon) with the purpose of the registration of the bioelectric muscle tension of selected muscles (soleus, tibialis anterior). A random group of senior subjects aged 60–70 years, who participate in regular fitness program involving the formation of psychomotor activities (once a week) and focusing primarily on the motor coordination and low-intensity aerobic performance capacity was subjected to 4 tasks involving postural assessment. The results of the study were concerned with the analysis of the correlation between the sEMG and COP parameters and confirm the initial hypothesis regarding the existence of a positive correlation between the level of muscular coactivation (soleus, tibialis anterior) and the scatter parameters of COP.

Keywords: Electromyography · Posturography · Age · Coordination

1 Introduction

Surface electromyography (sEMG) and posturography are some of the most useful tools applied in the assessment of the motor skills of humans. Described on the grounds of theory of the control and regulation of motor activities of humans, the motor system is based on reflexes occurring at the level of the spinal cord and along afferent and efferent pathways, thus involving a small degree of the conscience. This type of motor activity deals with the maintenance of the postural balance and walking activity. The motor habits in the form of automatism and voluntary movement (VM) are subjected to the control of the central nervous system (CNS) in particular in the motor cortex and cerebellum. The commands are carried from the decision centers initiate motor units (MU) through motoneurons, and result in the activation of skeletal muscles. The number of motor units activated gives the bioelectric muscle tension that can be expressed by means of the value of the sEMG signal [1]. In research, the measurement of bioelectric tension provides information about the level of the activity of the motor

W. P. Hunek and S. Paszkiel (Eds.): BCI 2018, AISC 720, pp. 23–29, 2018.
https://doi.org/10.1007/978-3-319-75025-5_3

units, which is correlated with the muscle strength. In addition, due to the timing applied to describe the excitement of particular muscles and muscle groups, it is possible to gain insight into the structure of the desired and pathological movement patterns. The approach that is described here can be applied in the analysis of the structure of the sEMG signal as the flow of information processes in combination with the feedback networks and outcomes perceptible in the form of postural control models [2]. In this sense, sEMG analysis forms a valuable tool as it can be applied in the teaching of sport technique, preventing injuries, and performing therapy applying the phenomenon of biofeedback [3]. The temporary loss of function, e.g. loss of lower extremity function is accompanied by a neuro-physiological process in which the neuronal representations are deteriorated in the motor control region of the cortex. In this sense, therapeutic strategy forms a psychomotor process by integrating the muscle functions with the CNS.

A systematic approach to this reveals the existence of interactions between the external stimulation and the stimuli originating from inside the human body. The internal excitation and the response to external stimuli are responsible for the control of the biological system based on feedback. The process of information exchange occurs constantly between the spinal cord and proprioceptors located inside tendons and muscles. This process is responsible for the spatial orientation, and maintenance of dynamic and static balance.

The assessment of the balance control is performed by the application of force plate posturography and the results are gained by analysis of the oscillations (sways) of the subjects' bodies based on knowledge of the center of pressure (COP) trajectory and measured in the frontal and sagittal plane as a result of the foot pressure exerted on a force plate. The adequate balance control in the healthy persons results from the set of sensory inputs from the vestibular system [4]. In this context, the authors decided to put the subjects to a testing procedure involving postural balance and proprioception tests. The tests of balance and visual control were performed with the subjects' open and closed eyes while the testing of the tactile sense distortion was undertaken on surface covered with a foam pad with a thickness of 10 cm.

The scope and objective of this paper is to report on concurrent and synchronized application of the Kistler platform for postural balance and electromyography analysis (sEMG Noraxon) with the purpose of the registration of the bioelectric muscle tension of selected muscles (soleus, tibialis anterior). From the review of the literature, we can conclude that there should be a positive correlation between the level of muscle co-activation (soleus, tibialis anterior) and the stabilogram parameters of COP.

The study applied a random group of senior subjects aged 60–70 years, who participate in regular fitness program involving the formation of psychomotor activities (once a week) and focusing primarily on the motor coordination and low-intensity aerobic performance capacity. In such exercise, we adopt the presumption that aging affects the processes of atrophy of the neural system, deterioration in functional capacities and a general neuro-muscular decline. These processes can be counteracted by a fitness program that focuses on aspects helping to provide well-being adequate to the age of the participants [5, 6].

2 Materials and Methods

The study involved a group of 53 aged 60–70 (height 160.3 ± 6.7, body mass 71.4 ± 13.6). The participants of the study were with provided information regarding the objective and course of the study and were asked to give legal consent to participate in the study. In addition, the scope and goal was approved by the Bioethics Committee of the Chamber of Physicians, so study was performed in accordance with the guidelines defined in the Helsinki Declaration for the conduct of clinical trials in humans.

The registrations of COP was performed by a force plate (Kistler type 9286AA, Winterthur, Switzerland), with a sampling frequency of 100 Hz, test duration: 30 s. The electromyographic study applied 16 channel sEMG signals (DTS type, Noraxon, Scottsdale, USA) recorded with the 16 bit resolution and a sampling frequency of 1500 Hz. The procedure of the registration of the bioelectric activity of the right and left tibialis anterior and soleus muscles followed SENIAM methodology [7, 8]. Prior to the procedure, the measurements spot on the body was prepared by removing hair so as to improve the electrical contact between the electrodes and the skin. The surface electrodes (Ag/AgCl) were situated on the belly of the muscle between the motion point and the tendon origin along the midline of the muscle. The signal processing and sEMG analysis applied NORAXON MR-XP 1.07 Master Edition software.

Prior to performing the projected tasks, each of the subjects was asked to move the foot to create the maximum dorsiflexion to gain the measurement of the maximum voluntary contraction (MVC) of the tibialis anterior muscle (TA) and then subjects performed plantar flexion so as to gain information regarding the MVC of the soleus muscle (SOL). The measurements were performed over a five-second voluntary contraction. The registered maximum bioelectric muscle activity was registered as a reference value applied in the normalization of the sEMG signal amplitude.

The major experiment involved the following four trials: (1) standing with open eyes on the plate (OO), (2) standing with closed eyes on the plate (OZ), (3) standing with open eyes on a foam pad placed on a force plate (OG), (4) standing with closed eyes on a foam pad placed on a force plate (ZG).

The subjects maintained the following position: 14° angle between the feet and 17 cm distance between the heels [9].

The instantaneous center of foot pressure (COP) were calculated from the components of forces of the registered plate response were separately analyzed in then medio-lateral (ML) and anterior-posterior planes (AP). The COP signal was applied for calculation of the linear and nonlinear parameters characterizing the postural control. The linear parameters include standard deviation of the time series (SD in mm) and mean velocity (MV w mm/s). The lower values of these parameters denote a more effective postural control. The entropy (SE, non-linear and dimensionless parameter) forms the measure of the irregularity and unpredictability of the time series. In the literature, this measure is associated with the amount of attention needed to perform a postural ask and level of automatism in performing this activity [10, 11].

The sEMG signals were smoothened by calculation of the root mean square (RMS) that was derived in the time window of 300 ms. The reference value of the

MVC was automatically derived in the time window of 1000 ms for which the mean value of the sEMG signal was the highest.

The coactivation of the muscles was calculated using the Falconer and Winter method [12] based on the formula: $CI = 2I_{ant}/I_{tot} \times 100\%$, where: I_{ant} is the magnitude of the antagonist muscle activity, and I_{tot} is the total muscle activity. The co-activation was derived separately for the left (CI L) and right (CI R) lower extremity.

The parameters of the COP and sEMG signals were subjected to the Shapiro-Wilk test for normality. The distributions of the analyzed variables were not found to be significant so the assumption of normality was fulfilled. For correlations between values of sEMG and COP signal parameters, Pearson r tests were performed. P-values ≤ 0.05 were considered as statistically significant. All statistics were performed using Statistica v.13.1 (StatSoft, Inc., OK, USA).

3 Results

The descriptive statistics with regard to all tasks performed by the subjects are summarized in Table 1, whereas the correlations are found in Table 2.

Table 1. Mean and standard deviations of COP and sEMG throughout all performed tasks.

	Task			
	OO	OG	OZ	ZG
ML				
SD (mm)	2.72 ± 1.08	4.86 ± 1.82	2.62 ± 0.84	6.31 ± 2.21
MV (mm/s)	5.69 ± 1.80	11.32 ± 3.36	6.39 ± 2.15	17.75 ± 7.10
SE	0.74 ± 0.23	0.62 ± 0.15	0.76 ± 0.17	0.62 ± 0.08
AP				
SD (mm)	4.53 ± 1.41	7.04 ± 2.67	4.72 ± 1.31	8.81 ± 2.87
MV (mm/s)	10.51 ± 4.11	14.80 ± 3.78	14.03 ± 7.56	26.23 ± 11.99
SE	0.75 ± 0.20	0.62 ± 0.15	0.80 ± 0.20	0.66 ± 0.11
CI R (%)	46.93 ± 28.99	28.89 ± 23.37	45.93 ± 28.57	44.07 ± 25.37
CI L (%)	48.88 ± 29.30	31.10 ± 22.37	49.54 ± 27.75	47.17 ± 26.32

The mean values of SD and MV COP are considerably greater in both planes for the task performed on a foam pad. A reverse relation was recorded for the case of the SE coefficient. Similar differences as the ones for SE can be noted for the value of CI, however, this happens only for the task performed with the subjects' open eyes.

A statistically significant correlation can be recorded only for the case of the dependence between MV AP and CI L parameters. The remaining correlations are low and statistically non-significant.

The plot with the exemplary courses of COP and sEMG for the right lower extremity during the task involving standing on a foam pad is presented in Fig. 1.

Table 2. Correlation coefficients for COP and sEMG parameters registered throughout particular tasks

	Task							
	OO		OG		OZ		ZG	
	CI R	CI L	CI R	CI L	CI R	CI L	CI R	CI L
SD ML	−0.148	−0.024	0.006	0.162	−0.166	−0.066	0.204	0.093
SD AP	−0.082	0.041	0.053	0.126	0.001	0.114	0.187	0.184
MV ML	−0.095	0.009	0.050	0.152	0.036	0.167	0.172	0.126
MV AP	−0.001	0.168	0.266	0.298*	−0.014	0.231	0.164	0.142
SE ML	0.081	−0.076	0.098	−0.084	0.208	0.119	0.015	0.128
SE AP	0.072	0.014	0.070	−0.047	0.042	0.024	0.046	0.041

*denotes statistically significant correlation coefficients $p < 0.05$

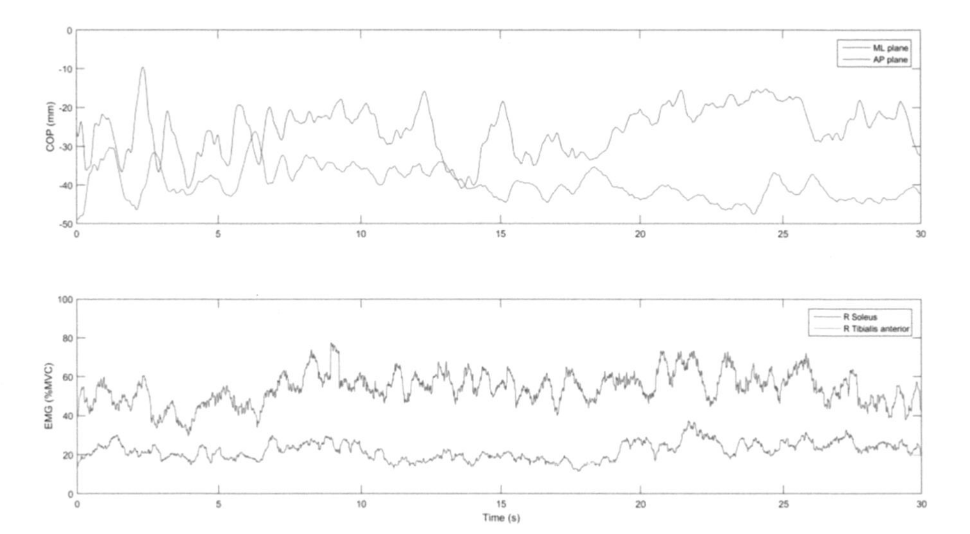

Fig. 1. Exemplary plot of sEMG and COP time series for foam pad with eyes open task.

4 Discussion

The objective in the current study involved the demonstration of the applicability of sEMG and posturography as the tools for the analysis of biosignals in the aging process in subjects in the post-production age, who keep regular activity by participating in the classes organized as part of the University of the Third Age. The results of the study were concerned with the analysis of the correlation between the sEMG and COP parameters and confirm the initial hypothesis regarding the existence of a positive correlation between the level of muscular coactivation (soleus, tibialis anterior) and the scatter parameters of COP. The statistically significant correlation between the coactivation of the muscles of the left lower extremity and the mean COP velocity in the anterior-posterior plane at a level of 0.298 was recorded only for the case of the task performed on a foam pad with open eyes.

The reports in the literature contain the results which differ considerably from the ones that were gained in the present study. The studies in [13, 14] gained a correlation between the coactivation and the magnitude of the oscillations equal to 0.278; however, the task involved standing with legs closed for only 10 s and coactivation that was analyzed throughout a three-second sample of sEMG. In a study reported by Kouzaki and Shinohara [15], a positive correlation was established between sEMG normalized in relation to MVC and the fluctuations of COP in the senior subjects. This study involved a comparison of the fluctuations of the force during a plantar bending with an equal intensity in the young and senior persons and their relations with fluctuations of COP in the standing still position. The results of this study demonstrate the variability of COP has a positive correlation with the coefficient of variation of force-time curves during plantar bending test of the feet. The seniors demonstrated an increased activity of antagonist muscles, which was found to relate to greater fluctuations of COP. The authors of this study conclude that this approach can be also applied to compare the results gained in healthy persons and the subjects with neurological disorders.

In a study performed by Donath et al. [16], an attempt was undertaken to determine the postutral stability in the senior and young subjects by analyzing the sEMG activity of the muscles (tibialis anterior, soleus, peroneus longus, gastrocnemius medialis) and COP displacement. This test applied a standing on both legs with eyes closed test and standing on one leg with eyes open. The authors in this study applied interval exercises with a high intensity performed on a treadmill between the tasks included in the testing procedure. The analysis involved the co-activation of the tibialis anterior and soleus muscles by application of the index ($CAI = 2 \times TA/TA + SOL$). The results demonstrate that an increased coactivation of the ankle joint occurs in the anterior-posterior plane only for test involving standing on one leg. In the group of seniors, a decreased postural control was observed accompanied by an increased bioelectric activity of the tibialis anterior muscle. In addition, this study demonstrated that the high-intensity exercise does not affect the results of the test.

The differences between the current study and the reports from the literature could be attributable to the differences in the testing protocol, use of analytical tools, test duration and position of the feet throughout the tasks. The results of the experiments demonstrate a good level of applicability of the presented approach involving testing balance control in senior subject.

References

1. Latash, M.: Neurophysiological Basis of Movement, 2nd edn. Human Kinetics, Champaign (2008)
2. Borysiuk, Z., Waskiewicz, Z.: Information processes, stimulation and perceptual training in fencing. J. Hum. Kinet. **19**, 63–82 (2008). https://doi.org/10.2478/v10078-008-0005-y
3. Balkó, Š., Borysiuk, Z., Balkó, I., Špulák, D.: The influence of different performance level of fencers on muscular coordination and reaction time during the fencing lunge. Arch. Budo **12**, 49–59 (2016)

4. Derlich, M., Krecisz, K., Kuczyński, M.: Attention demand and postural control in children with hearing deficit. Res. Dev. Disabil. **32**, 1808–1813 (2011). https://doi.org/10.1016/j.ridd.2011.03.009
5. Sacha, J., Sacha, M., Soboń, J., Borysiuk, Z., Feusette, P.: Is it time to begin a public campaign concerning frailty and pre-frailty? A review article. Front. Physiol. **8**, 484 (2017). https://doi.org/10.3389/fphys.2017.00484
6. Puciato, D., Borysiuk, Z., Rozpara, M.: Quality of life and physical activity in an older working-age population. Clin. Interv. Aging **12**, 1627–1634 (2017). https://doi.org/10.2147/CIA.S144045
7. Merletti, R., Parker, P.: Electromyography: Physiology, Engineering, and Non-invasive Applications. Wiley-IEEE Press, Piscataway (2004)
8. Mika, A., Oleksy, Ł., Kielnar, R., Wodka-Natkaniec, E., Twardowska, M., Kamiński, K., et al.: Comparison of two different modes of active recovery on muscles performance after fatiguing exercise in mountain canoeist and football players. PLoS One **11**, e0164216 (2016). https://doi.org/10.1371/journal.pone.0164216
9. McIlroy, W.E., Maki, B.E.: Preferred placement of the feet during quiet stance: development of a standardized foot placement for balance testing. Clin. Biomech. (Bristol, Avon) **12**, 66–70 (1997). http://www.ncbi.nlm.nih.gov/pubmed/11415674
10. Donker, S.F., Ledebt, A., Roerdink, M., Savelsbergh, G.J.P., Beek, P.J.: Children with cerebral palsy exhibit greater and more regular postural sway than typically developing children. Exp. Brain Res. **184**, 363–370 (2008). https://doi.org/10.1007/s00221-007-1105-y
11. Bieć, E., Zima, J., Wójtowicz, D., Wojciechowska-Maszkowska, B., Kręcisz, K., Kuczyński, M.: Postural stability in young adults with down syndrome in challenging conditions. PLoS One **9**, e94247 (2014)
12. Falconer, K., Winter, D.A.: Quantitative assessment of co-contraction at the ankle joint in walking. Electromyogr. Clin. Neurophysiol. **25**, 135–149 (1985)
13. Nagai, K., Yamada, M., Uemura, K., Yamada, Y., Ichihashi, N., Tsuboyama, T.: Differences in muscle coactivation during postural control between healthy older and young adults. Arch. Gerontol. Geriatr. **53**, 338–343 (2011). https://doi.org/10.1016/j.archger.2011.01.003
14. Nagai, K., Yamada, M., Mori, S., Tanaka, B., Uemura, K., Aoyama, T., et al.: Effect of the muscle coactivation during quiet standing on dynamic postural control in older adults. Arch. Gerontol. Geriatr. **56**, 129–133 (2013). https://doi.org/10.1016/j.archger.2012.08.009
15. Kouzaki, M., Shinohara, M.: Steadiness in plantar flexor muscles and its relation to postural sway in young and elderly adults. Muscle Nerve **42**, 78–87 (2010). https://doi.org/10.1002/mus.21599
16. Donath, L., Kurz, E., Roth, R., Zahner, L., Faude, O.: Different ankle muscle coordination patterns and co-activation during quiet stance between young adults and seniors do not change after a bout of high intensity training. BMC Geriatr. **15**, 19 (2015). https://doi.org/10.1186/s12877-015-0017-0

The Possibilities of Using BCI Technology in Biomedical Engineering

Dariusz Man(✉) and Ryszard Olchawa

Institute of Physics, Opole University, Oleska 48, 45-052 Opole, Poland
{dariusz.man,rolch}@uni.opole.pl
http://fizyka.uni.opole.pl

Abstract. The paper presents capabilities of building devices dedicated for persons with heavy mobility dysfunction and indicates the role of interfaces connecting brain with computer (Brain Computer Interface, BCI). Impulses coming from closing eyes, clenching teeth, and tongue movement were proposed as optimal in controlling the applications that manage executable systems. A group of electrodes giving a strong electric signal characteristic for the activity were designated and on the basis of conducted research a proposition of a scientific project concerning building of supporting devices for persons with heavy mobility dysfunction was presented.

Keywords: Brain Computer Interface · EEG · Biomedical engineering

1 Introduction

In the past years it has been possible to notice the rapid growth of digital technology which utilizes action potential of the brain. Technological support of neurophysiology allowed relatively simple way to register brain waves which in consequence led to creation of integrated BCI (Brain Computer Interface) interfaces [1–3] used e.g. in EEG biofeedback sets. This technology found its use in managing adequately prepared applications for PC computers, smartphones, and tablets. It is also possible to use this technology to control everyday electronics as well as technologically advanced devices by persons with severe neurological diseases which is important in rehabilitation process and improvement of quality of life for such persons [3–7]. It is impossible to overestimate the potential of BCI technology in biocybernetics and biomedical technology. Our goal was to find out whether cheap and effective ways of implementing BCI technology for application and device control that would support persons with movement dysfunction exist. Signal analysis shows that strong, characteristic electric impulses exist and are located in specific brain areas. The shape and amplitude of impulses were similar in every case. The results also show a possibility to create a universal algorithm for technical devices, PC computers and mobile devices applications control, supporting the use by multiple users.

W. P. Hunek and S. Paszkiel (Eds.): BCI 2018, AISC 720, pp. 30–37, 2018.
https://doi.org/10.1007/978-3-319-75025-5_4

2 Materials and Methods

Action potential mapping sets allowing monitoring and recording complete EEG image were used during the research. EMOTIV brand devices used during the research consisted of 14 channel helmets (AF3, F7, F3, FC5,T, P7,O1, O2, P8, T8, FC6, F4, F8, AF4) with two reference electrodes (P3/P4) in CMS/DRL configuration, BCI interface, and PC software. Basic parameters of measuring system used in research were: 128 or 256 SPS sampling frequency, 14 or 16 bit resolution, built-in Sinc 5-row filter. Figure 1 depicts electrode arrangement on the head of test subject.

In order to gather the characteristic EEG impulses, 20 persons (10 men and 10 women) of various ages (age ranging from 22 to 81) were tested During testing a special algorithm was used which consisted of sequence of commands that the test subjects were to follow. This allowed the precise time measurement of brain responses to requested psychomotoric activities. The conducted research shows that the best results with high repetitiveness rate and amplitude were observed during the device reacting to blinking of eyes, clenching of teeth, and tongue movements. Figure 2 depicts exemplary EEG spectrum during the test in which test subjects were blinking with the left and right eyes simultaneously. Test subjects also performed two blinking cycles of 9 activities with 0,5 s interval. As presented on the diagram a strong and repetitive signal was registered by AF3 (left eye) and AF4 (right eye) electrodes.

F7, F3 (left eye) and F4,F8 (right eye) electrodes also registered a slightly weaker signal. Subtle changes can also be observed on FC5 and FC6 electrodes. The remaining electrodes located further away from the eyes did not register action potential of the blinking activity.

Figure 3 depicts exemplary EEG signal image of individual blinking activity of the left and right eye. Test subject performed series of 3 blinks in a 2 s interval. Diagram shows strong signal registered for the left eye with AF3 and F7 electrodes, and AF4 and F8 electrodes respectively.

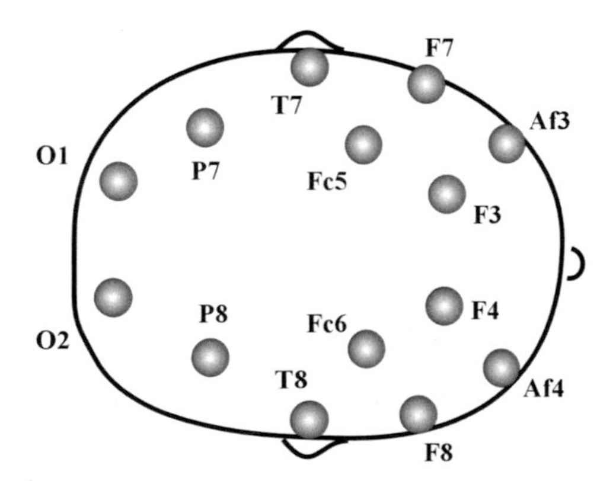

Fig. 1. Schematic of electrodes arrangement and identification on the head of test subject.

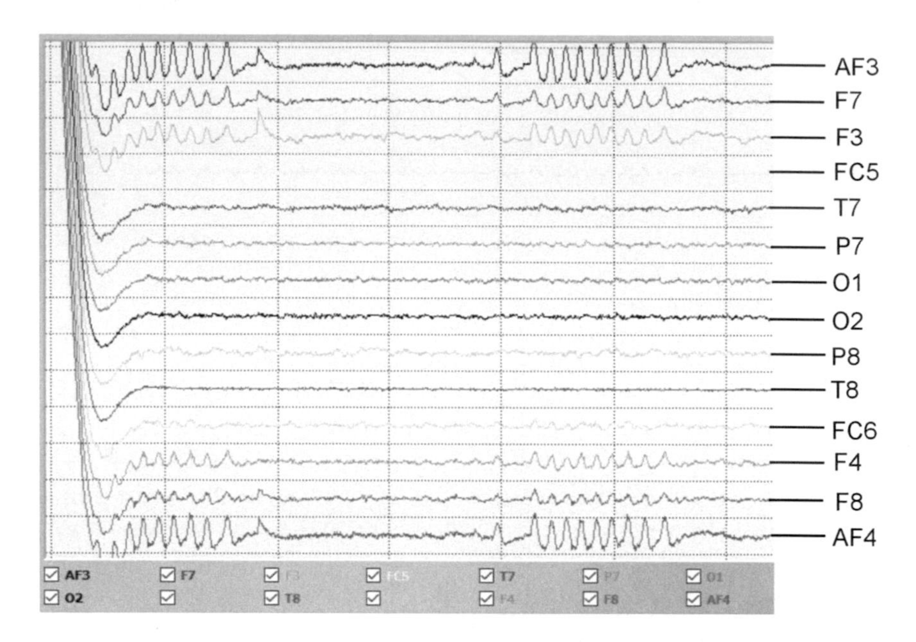

Fig. 2. EEG signal registered during simultaneous left and right eye blinking activity. EMOTIV brand 14 electrode EEG helmet.

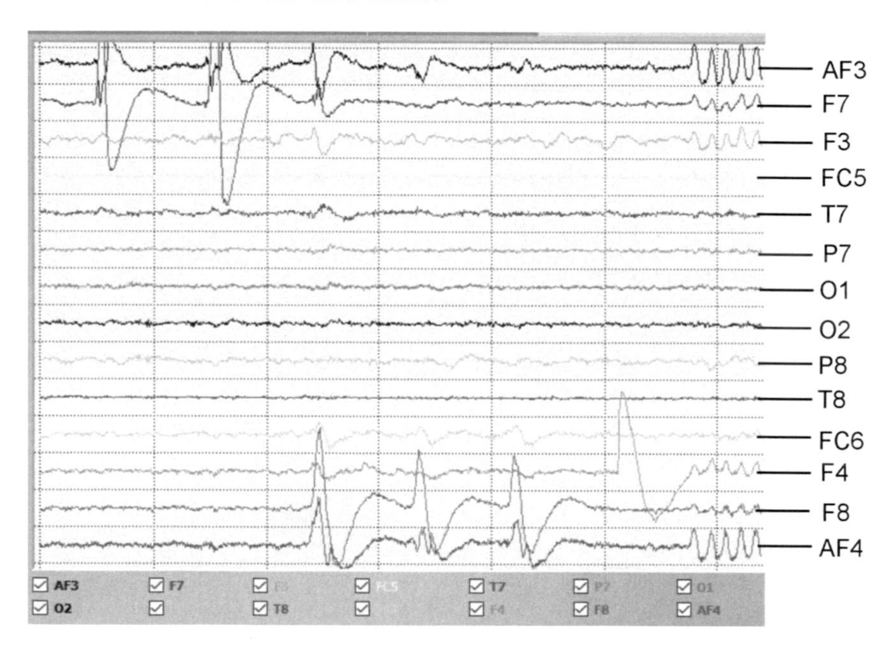

Fig. 3. EEG signal registered during individual blinking activity with left and right eye. EMOTIV brand 14 electrode EEG helmet.

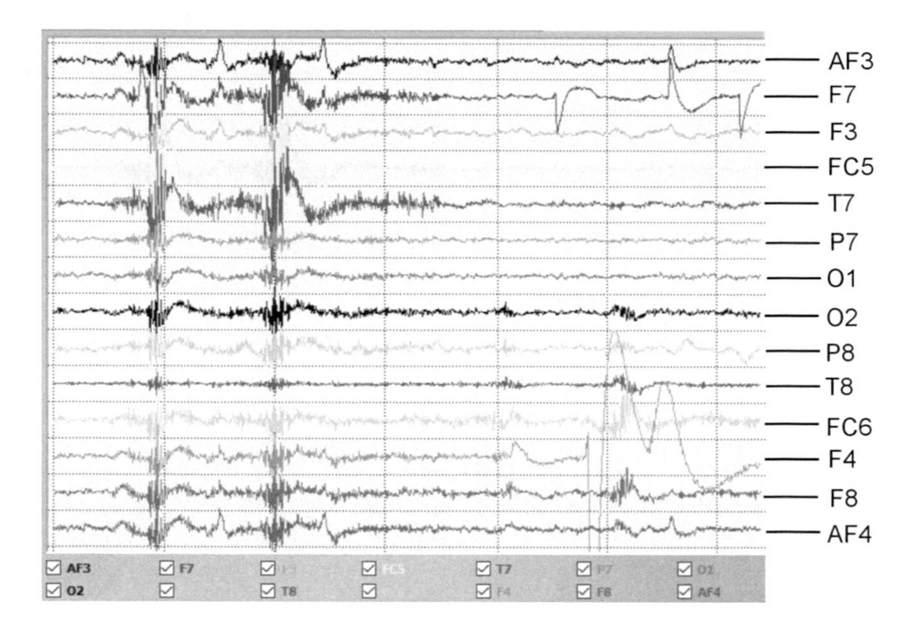

Fig. 4. EEG signal registered during teeth clenching activity. EMOTIV brand 14 electrode EEG helmet.

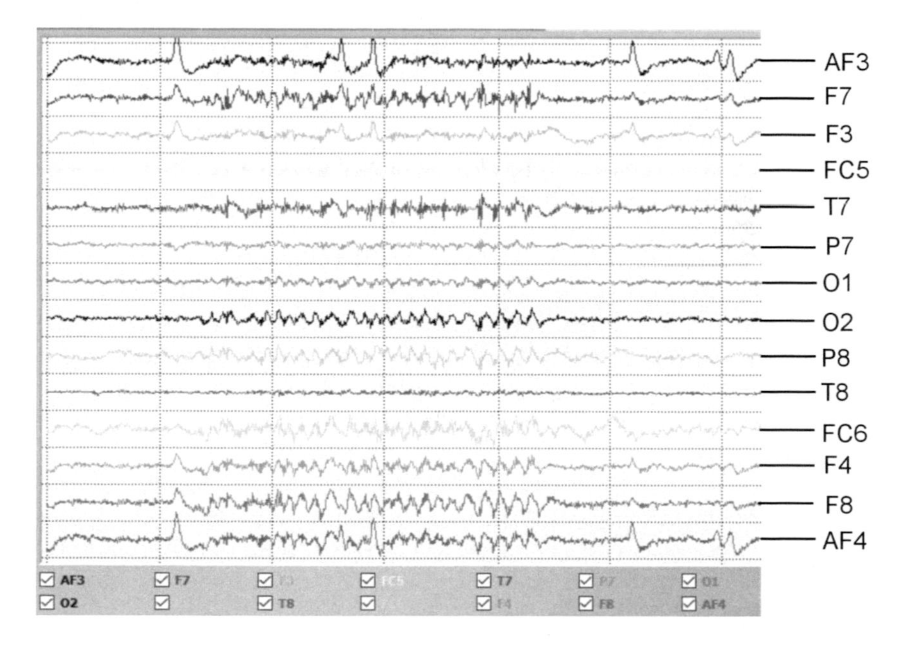

Fig. 5. EEG signal registered during tongue movement activity. EMOTIV brand 14 electrode EEG helmet.

Figure 4, depicts exemplary EEG signal image of teeth clenching activity. Test subject performed series of 2 clenches in a 2 s interval. A clear signal was registered with almost all electrodes which may be surprising particularly in case of O1 and O2, and P7 and P8 electrodes.

Figure 5 depicts exemplary EEG signal image of tongue movement activity. Test subject performed series of several movements. Similar to teeth clenching, signals were registered by almost all electrodes, however with much weaker amplitude.

3 Data Processing

EMOTIV brand software allows for data registry of particular electrode together with the clock signal information. The file format is based on the European Data Format (EDF) widely used for exchange and storage of multichannel biological and physical signals. However, data format used by EMOTIV do not follow requirements of the EDF standard which makes it impossible to directly open the files with software other than EMOTIV. On the other hand, the exceptions are minuscule enough to make it possible to individually create software that allows for data extraction for further processing.

The primary goal of data processing is extraction and identification of signals connected to particular action potentials presented in Figures 2, 3, 4 and 5. High amplitude of such signals and strong connection to specific electrodes allows for reliably identification and assignment of action potentials.

The first phase of registered data processing is noise elimination in signals $f_e(t_n)$, with e standing for electrode, and tn for sampling time. To do this, discrete Fourier transform is used $g_e(\omega_k) = DFT[f_e(t_n)]$, and after eliminating components with higher frequencies $\omega_k = 0$ for $\omega_k > \omega_h$ a reverse Fourier transform is performed. FFT algorithms allow for performing this activity in real time using ARM, and even ARV, energy-saving microcontrollers. Drift of the constant component observable in the beginning phase of registering in Fig. 2 can be easily eliminated by subtracting registered potential on reference electrode: $f_e(t_n) := f_e(t_n) - f_r(t_n)$, with $f_r(t_n)$ being the signal on the reference electrode. After the initial preparation signals undergo threshold analysis that allows for identification of high amplitude signal sequences coming from action potentials. In order to increase reliability during the processing, the signals undergo the process of normalization by scaling to established amplitude. Signals normalized in this way are then compared with a template in order to eliminate signals coming from different sources. The measure δ of matching to a template, equals to squared average deviation between the pattern $w(t_n)$ and the signal $f_e(t_n)$ which can be written as

$$\delta = \langle (f_e(t_n t_p) - w(t_n))^2 \rangle_{t_n}.$$

Parameter t_p of this equation determines the signal time delay in relation to the pattern and is set in a way that makes δ take the minimal value. This approach enables reliable signal identification and precise time mark of the registered event. Thanks to good time determination it is possible to precisely inspect time sequences. The presented algorithm of signal processing is possible to implement on miniature microcontrollers without the use of PC computer. Further interpretation of various combinations and sequences of activities performed with the use of eyes, teeth clenching or tongue movements enables enormous possibilities of device control. Such controls can be managed by the same microcontroller through the use of adequate states on its external ports. Imagine generating a text by moving the cursor over the virtual keyboard or with some experience by sequences of blinks basing on the Morse code.

4 Discussion of Results

On the basis of conducted research it is possible to state that activities connected with blinking and teeth clenching are accompanied by clear and recurring impulses. This makes it possible to use these currents to control supporting devices by persons with movement dysfunction. EEG signal analysis allows pointing the optimal placement of electrodes on the patient's head. Diagrams Figs. 2, 3 and 4 show that headband with two electrodes placed in the same location as electrodes AF3 and AF4 in EEG Helmet is just enough. A two-electrode headband is also enough in the case of teeth clenching. This solution considerably decreases the price of the apparatus and the inconvenience of use for the patient which in return increases the comfort of use of technical devices. In our opinion there is a real possibility to create a set of supporting devices for persons with movement dysfunction. In case of measurement necessity of currents coming from tongue movements, more complex algorithms are needed due to relatively weak impulses with heterogeneous shapes [8–10]. Measurement headband will have to be equipped with 4 electrodes.

5 Summary

The solution we propose in Fig. 6 assumes using electronic components to build a headband dedicated to control the rehabilitation bed, wheelchair, and applications that enable the use of editors and web browsers. Initial works show the possibility of complete execution of the project with building prototypes for about 200,000 which is financially attractive considering the nature of medical projects.

Planned realization date of the project is in 3 years from acquiring the funding. In the first year we plan wide research on various patient groups in agreement with neurological hospitals. The phase of executable algorithm and EEG headband creation will begin after data collection (minimum of 100 persons) and processing (second year of the project). During the third year the first prototypes will be created, rehabilitation beds and wheelchairs will be adjusted to EEG communication, and PC computers and mobile devices communication software will be developed.

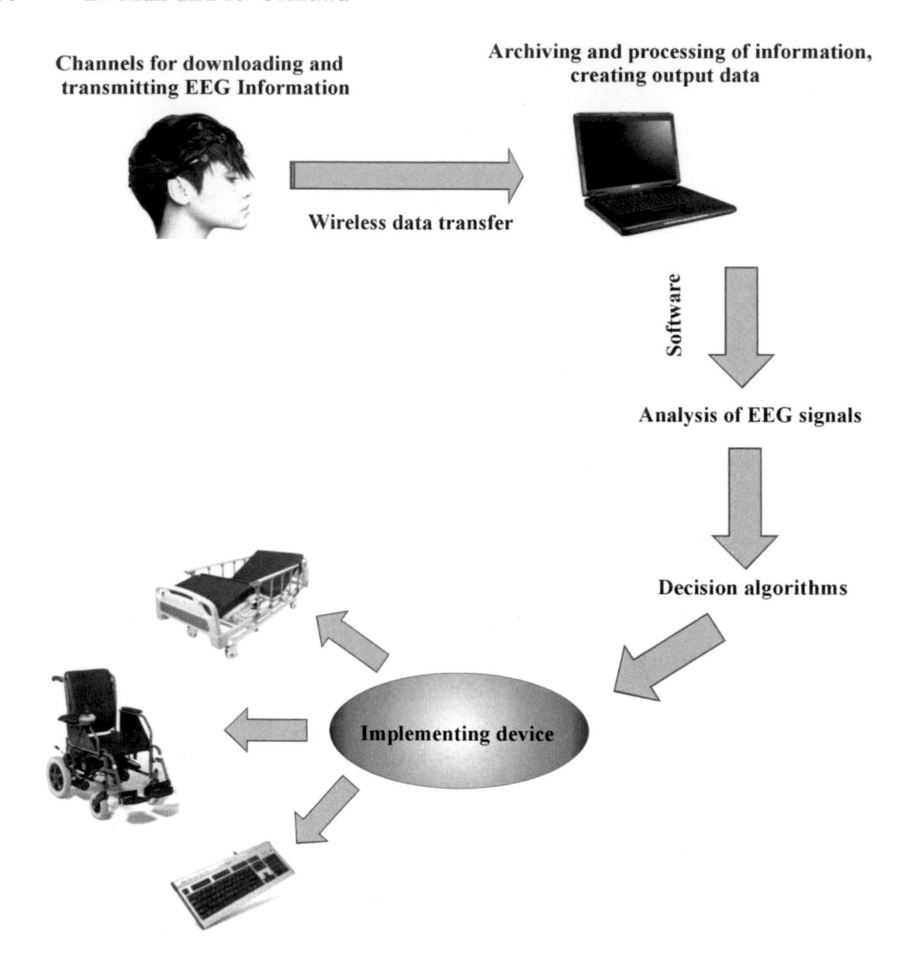

Fig. 6. Schematic representation of the project's idea of support for persons with movement dysfunction

References

1. Nicolas-Alonso, L.F., Gomez-Gil, J.: Brain computer interfaces, a review. Sensors (Basel). **12**(2), 1211–1279 (2012)
2. Shih, J.J., Krusienski, D.J., Wolpaw, J.R.: Brain-computer interfaces in medicine. Mayo Clin. Proc. **87**(3), 268–279 (2012)
3. Looney, D., Kidmose, P., Mandic, P.: Ear-EEG: user-centered and wearable BCI. Brain-Comput. Interface Res. Biosyst. Biorobot. **6**, 41–50 (2014)
4. Hassanien, A.E., Azar, A.T. (eds.): Brain-Computer Interfaces. Intelligent Systems Reference Library, vol. 74. Springer, Heidelberg (2015)
5. Graimann, B., Allison, B., Pfurtscheller, G. (eds.): Brain-Computer Interfaces. The Frontiers Collection. Springer, Heidelberg (2010)
6. Alwasiti, H., Aris, I., Jantan, A.A.: Brain computer interface design and applications: challenges and future. World Appl. Sci. J. **11**(7), 819–825 (2010)

7. Fazel-Rezai, R. (ed.): Brain-Computer Interface Systems - Recent Progress and Future Prospects, Chap. 11, pp. 37–39. InTech (2013)
8. Kameswara, T., Rajyalakshmi, M., Prasad, T.: An exploration on brain computer interface and its recent trends. Int. J. Adv. Res. Artif. Intell. (IJARI) **1**(8), 17–22 (2013)
9. Durka, P.J.: Matching Pursuit and Unification in EEG analysis. Engineering in Medicine and Biology. Artech House, Norwood (2007)
10. Durka, P.J., Matysiak, A., Montes, E.M., Valdes-Sosa, P., Blinowska, K.J.: Multichannel matching pursuit and EEG inverse solutions. J. Neurosci. Methods **148**(1), 49–59 (2005)

Brain Biophysics: Perception, Consciousness, Creativity. Brain Computer Interface (BCI)

Dariusz Man(✉) and Ryszard Olchawa

Institute of Physics, Opole University,
Oleska 48, 45-052 Opole, Poland
{dariusz.man,rolch}@uni.opole.pl
http://fizyka.uni.opole.pl

Abstract. The paper presents connections between perception, awareness, and creativity from the biophysical point of view. Attention was drawn to human senses' limitations and their influence on cognition. The role of interfaces connecting brain with computer and particular role of Brain Computer Interface (BCI) are indicated which the authors believe will be the next stage of human brain supporting technology evolution. It will enable the growth of perception, awareness, and creativity, and consequently lead to social development.

Keywords: Brain biophysics · Perception biophysics
Positive feedback

1 Introduction

Human brain is the most complex biophysical structure that the science tries to describe. Despite multiple attempts to recreate the essence of brain, it's functioning is still a mystery. It may be tied to large amount of it's building elements [1] or not yet discovered mechanisms of this complex environment. By analyzing brain as a physical structure it is possible to distinguish high number of logical elements (neurons) above all. This number is estimated to be 10^{11}, and due to numerous synapses (about seven thousand per neuron) results in 7×10^{14} functional connections. The human brain has been estimated to contain approximately 100 trillion synapses [2]. Such amount equals to about 88 TB of computer data storage and nowadays does not seem shocking. The estimated connection amount of 7×10^{14} is achieved in an assumption that every synapse is permanently tied to one other neuron. In this case, a selected synapse can be attributed with logical value of 1 if it is connected to a selected neuron or logical value of 0 if it is not connected. However, if we consider the dynamic nature of synapse and neuron connections and assume it may connect with any other neuron, it is possible to attribute synapse with logical value of 0 if it is not connected with any neuron or logical values from 1 to $(10^{11} - 1)$ in case of connection with one of $(10^{11} - 1)$ remaining neurons. This results in a drastic increase in possible functional connections configurations. Excluding configurations in which

W. P. Hunek and S. Paszkiel (Eds.): BCI 2018, AISC 720, pp. 38–44, 2018.
https://doi.org/10.1007/978-3-319-75025-5_5

two synapses of the same neuron connect with the same one from the remaining neurons, the number of connections of one neuron with others equals to number of combinations without repetitions given by Newton binomial series

$$\binom{10^{11}-1}{7\times 10^3}.$$

Total number of neuron connection configurations can be given by equation

$$\frac{\binom{10^{11}-1}{7\times 10^3}\times 10^{11}}{2},$$

with division by two allowing to avoid counting the same connection between neurons twice. The approximate value of this equation is 10^{53132}. Each of the connections forms a logical path that can be responsible for a specific action [3]. Synapses may be in one of two states: conduction, that is conducting the neural impulse (logical 1); or no conduction, that is blocking the neural impulse (logical 0). Therefore, on the basic level the brain functions in a binary system, similar to processors in our computers. Each stimulated neuron is accompanied by electric current which in turn induces electromagnetic field. Similar to computers, brain functioning is accompanied by a specific background of varying electric and electromagnetic potentials. These potentials can be registered in the form of EEG signal (electroencephalograph). Richard Catona was the first man to register these potentials on open brains of rabbits and monkeys. The results together with experiment description were published in British Medical Journal in 1875. Many years later, in 1924 psychiatrist Hans Berger conducted research on EEG use on people and published his results in 1929 [4]. However, EEG was fully approved by scientific community only in 1937. Currently, encephalography is one of routine medical diagnosis methods used by neurologists and psychiatrists despite its nature still not being sufficiently clarified. EEG measurement technology evolved together with development of electronics, physics, and neurology. Microelectronics and digital technology in particular accelerated its development by creating modern tools and opening new areas of research for scientists. One of such areas is brain-computer-interface (BCI) which allows direct brain control of adequately prepared applications [5–7]. It appears that researches on brain functioning enter a new phase and leave medical practices to become subjects of physics and engineering studies. This is an obvious success as the complex nature of the problem requires involvement of researchers of many scientific disciplines.

2 Perception

Perception is organization, identification, and interpretation of sensory information coming to brain through sensory organs in order to understand the environment [8]. On this basis we build ideas - a model of the world that surrounds us. In reality, it is subjective and far incomplete idea. We will now approach this problem from the physicist's perspective instead of the psychologist's. The first thing to notice is tremendous limitations of sensory organs which means that we miss an information, or rather a really big amount of it, in relation to what our senses can register and process. Our brains acquire about 90% of information from two senses only: sight (about 80%) and hearing (about 10%). The remaining 10% of information is distributed between smell, touch, and taste senses Fig. 1.

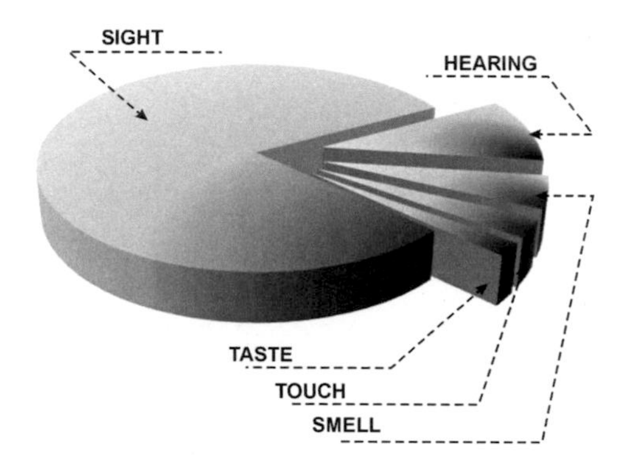

Fig. 1. The percentage schema input form of information acquired by the human senses: sight, hearing, smell, touch, taste.

The idea of reality built by brain is therefore mostly based on sight. We will now explore the physical capabilities of this detector which converts electromagnetic energy into nerve impulses (electric current). Eye construction ensures good processing of quanta of the energy E ($E = h\ \ni$: h - Planck's constant, $\ni$ - frequency) of the wave length scope from 380 nm to 740 nm [9,10]. However, the multitude of information that surrounds our brains is placed in scope of energy ranging from several thousand kilometers to a ten thousandth of a picometer of wave length Table 1.

For this reason majority of information in form of electromagnetic radiation that reaches the sight organ is invisible. The Universe and nature that surrounds us constantly sends out information about its energetic state at the speed of light but most of it eludes our perception. Similar to this situation is a problem of information acquisition via the hearing organ. Sound wave travels the air at the speed of about 340 m/s and our ears can process acoustic waves ranging from 20 Hz to 20,000 Hz (cycles per second) [11] which equals wave lengths from 17 m to 2 cm, Table 2. The reality is far worse as majority of us cannot register the full hearing range. While it is true that everyday communication (human speech) requires much narrower frequency range: 130–1000 Hz for women, 65–500 Hz for men; the world that surrounds us emits acoustic waves ranging from a fraction to millions Hz. A sound that is lower in frequency than 20 Hz is infrasound, and a sound above 20 kHz ultrasound. Many natural phenomena including water waves, wind, wing movement of birds, fan centrifugation, etc. generate infrasounds, often high powered, that may have negative influence on human organism. On the other hand phenomena including surface tension of

Table 1. Summary of parameters of electromagnetic waves (frequency and length) for various types of waves.

Types of electromagnetic waves. Sources	Frequency [Hz]	Wavelength [m]
Power engineering: Variable currents and pulse	1 to 10^2	300000×10^3 to 3000×10^3
Corded phones	10^2 to 10^4	3000×10^3 to 30×10^3
Hertz's waves	10^4 to 10^{13}	30×10^3 to 0.03×10^{-3}
Radio waves:		
Long waves	1.5×10^5 to 3×10^5	2000 to 1000
Medium waves	0.5×10^6 to 2×10^6	600 to 150
Short waves	0.6×10^7 to 2×10^7	50 to 15
Ultrafast waves	0.2×10^8 to 3×10^8	15 to 1
Microwaves: Radar, Technologies military	3×10^8 to 10^{13}	1 to 0.03×10^{-3}
Infrared: matter of temperature higher up 0K	10^{12} to 4×10^{14}	0.03×10^{-3} to 790×10^{-9}
Visible light: Sun	4×10^{14} to 8×10^{14}	790 to 390×10^{-9}
UV: sun, electric arc	8×10^{14} to 3×10^{16}	390 to 5×10^{-9}
Radiation of character quantum		
Roentgen's rays	3×10^{16} to 3×10^{20}	10000 to 1×10^{-12}
Rays γ	10^{18} to 10^{22}	300 to 0.03×10^{-12}
Cosmic rays	10^{22} to 10^{24}	0.03 to 0.0003×10^{-12}

Table 2. Summary of parameters of acoustic waves in the human hearing range. The range most frequently used in voice communication is marked in bold.

Frequency [Hz]	Length [m]
20	17
25	13.60
40	8.50
80	**4.25**
160	**2.125**
320	**1.062**
500	**0.68**
1000	**0.34**
2000	0.17
4000	0.085
10000	0.034
20000	0.017

crystal structures (e.g. rocks, steel, glass), piezoelectric effects, and echolocation in animals (e.g. bats, dolphins), generate ultrasounds. Unfortunately, this kind of information is unavailable for human mind. The world we build in our imagination is imperfect and perhaps even in certain cases entirely wrong. Therefore, how are we to perceive nature if our perception is flawed. Correct stimuli identification and interpretation by our brains is a completely different matter (we will leave this task for psychologists and psychiatrists).

3 Awareness

State or ability of being aware that is condition of thoughts, feelings, and will as well as accompanying phenomena; recognition of own deeds and feelings by a subject [12]. If awareness is mind's ability to reflect "objective reality" and constitutes the highest level of psychological development, we must consider what decides and builds awareness. It would appear that there are five distinct components that influence the level of. The first is perception which is responsible for gathering the information. Because the information must be processed and correctly allocated, knowledge and experience are necessary. In order to draw conclusions and take optimal decisions based on the processed informations we use intelligence. Our decisions and assessment are further verified by ethics. Therefore, perception, intelligence, knowledge, experience and ethics Fig. 2 model our awareness.

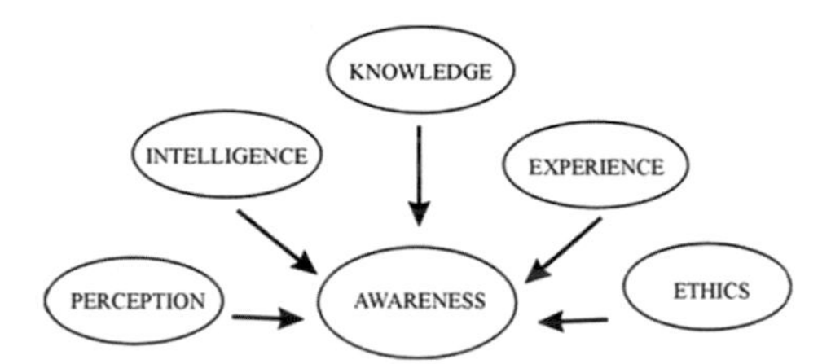

Fig. 2. Elements influencing the development of awareness.

It would appear that broadening the area of our knowledge through studying and perception by e.g. using certain interfaces we are able to influence our awareness. Physics, and particularly biophysics and BCI technology, play an important role here which is broadening the electromagnetic and acoustic waves registering and processing capabilities to an area currently unavailable for our senses. This will create entirely new cognitive capacity, particularly when linked directly to brain and will most probably have a significant influence on our awareness.

4 Creativity

Currently, there are a few more or less extensive definitions of creativity. The word itself is derived from Latin creatus which means creative. Therefore, creativity is a process which leads to emergence of new solutions, exactly

"Over the course of the last decade, however, we seem to have reached a general agreement that creativity involves the production of novel, useful products" Michael Mumford 2003 [13]. Social development is the effect of creativity which we owe to the growing awareness and the rapid human development during past decades is amazing. In the beginning of the XX century horses were the basic means of transportation, a solution that lasting for thousands of years. Half a century age text written on paper was the basic mean of information and nowadays thanks to digital technology we have a variety of options. In 1969 people landed on the Moon and there are thousands of satellites over our heads including the International Space Station. We are now timidly begin the era of space exploration, however what we have achieved as a species for the last hundred years can be considered a miracle when compared to everything our civilization achieved during thousands of years. There can be only one diagnosis of this enormous developmental acceleration - considerable increase in sciences' development, physics and chemistry in particular. This development was made possible thanks to the discovery and use of new devices that broadened our perception which became a positive feedback Fig. 3. Creativity provides the means to broaden perception which in consequence leads to the development of awareness and creativity.

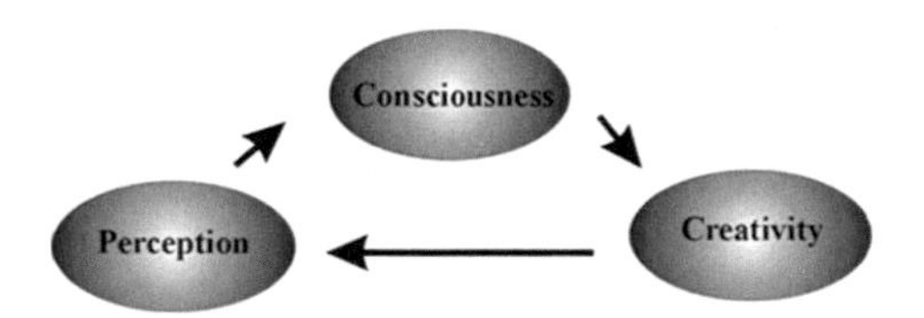

Fig. 3. Diagram of positive feedback reinforcing development of civilization.

5 Conclusion

It would appear that humanity stands before another phase of development acceleration. Technology allows significant broadening of cognition. Interfaces that connect our brains with computers and mediate the use of machines allow exploration of entirely new areas. Imagine a device that processes electromagnetic wave lengths ranging from radio waves to gamma waves in a way understandable for human brain instead of eyes with a limited functionality. Additionally, such device could increase the optic sensitivity and allowing magnification in any range (physically possible). Imagine a wide range interface that processes sounds from hearing range as well as infra- and ultrasounds instead of ears. This would certainly change our perception of the world and introduce the new area of awareness. In order to achieve it, we must further develop the BCI technology and create safe and non-invasive Brain Computer Interface. We must once again look at the problem from the biophysics' perspective to fully understand new capabilities that could appear thanks to this technology. The so-called reality is in fact a virtual world based on the acquired data and created by the neurons of our brains (this problem was discussed in detail during a lecture cycle in Opole University Institute of Physics and I and II Konferencja Mózg Komputer (BCI Conference) in Opole University of Technology and described [14]). The current state and perception capabilities of our main senses are surprisingly weak. If we assume that length of electromagnetic waves ranges from about 3×10^{-16} m to 3×10^{8} m in nature (24 orders of magnitude), our sight registers barely 10^{-9}% of available information. What does that mean? It is as if one was to make an opinion on the content of a library that contains a billion (109) books after reading a single book! Food for thought. In case of hearing, we are able to only register about 2×10^{-2}% of all acoustic information. Our main senses do not allow for reasonable data collection about the Universe. Here physics comes to our aid once again. Microscopes, telescopes, and spectroscopes built by scientists further increase the capabilities of sight. However, this knowledge is available to a small group of people. This, coupled with growing specialization in sciences often makes it impossible to take a holistic approach to problems. This is why creating new tools, most importantly the BCI, appears to be a necessity on one hand and on the other a natural process of creativity evolution of our civilization.

References

1. Pelvig, D.P., Pakkenberg, H., Stark, A.K., Pakkenberg, B.: Neocortical glial cell numbers in human brains. J. Neurobiol. **29**(11), 1754–1762 (2008)
2. Williams, R.W., Herrup, K.: The control of neuron number. Ann. Rev. Neurosci. **11**, 423–453 (1988)
3. Shepherd, G.M.: Introduction to synaptic circuits. In: The Synaptic Organization of the Brain. Oxford University Press, New York (2004)
4. Haas, L.F.: Hans Berger (1873–1941), Richard Caton (1842–1926) and electroencephalography. J. Neurol. Neurosurg. Psychiatr. **74**, 9 (2003)
5. Nicolas-Alonso, L.F., Gomez-Gil, J.: Brain computer interfaces, a review. Sensors (Basel) **12**(2), 1211–1279 (2012)
6. Shih, J.J., Krusienski, D.J., Wolpaw, J.R.: Brain-computer interfaces in medicine. Mayo Clin. Proc. **87**(3), 268–279 (2012)
7. Looney, D., Kidmose, P., Mandic, P.: Ear-EEG: user-centered and wearable BCI. Brain-Comput. Interface Res. Biosyst. Biorobot. **6**, 41–50 (2014)
8. Bernstein, D.A.: Essentials of Psychology, 5th edn, pp. 123–124. Wadsworth Publishing, Boston (2010)
9. Live Science: https://www.livescience.com/50678-visible-light.html
10. Gollisch, T., Meister, M.: Eye smarter than scientists believed: neural computations in circuits of the retina. Neuron **65**(2), 150–164 (2010)
11. The Physics Factbook. https://hypertextbook.com/facts/2003/ChrisDAmbrose.shtml
12. Kopaliski, W.: Słownik wyrazw obcych i zwrotow obcojezycznych, p. 389. Wiedza Powszechna, Warszawa (1967)
13. Mumford, M.D.: Where have we been, where are we going? taking stock in creativity research. Creat. Res. J. **15**, 107–120 (2003)
14. Man, D.: Biofizyka mozgu - Czy istnieje globalna swiadomosc? Wspolczesne problemy w zakresie inżynierii biomedycznej i neuronauk, Opole 7–14 (2016)

The Neglected Problem of the Neurofeedback Learning (In)Ability

Rafał Łukasz Szewczyk[1(✉)], Marta Ratomska[2], and Marta Jaśkiewicz[2]

[1] Department of Cognitive Psychology, SWPS University of Social Sciences and Humanities, Warsaw, Poland
rafal.lukasz.szewczyk@gmail.com
[2] Department of Experimental Psychology, The John Paul II Catholic University of Lublin, Lublin, Poland

Abstract. Neurofeedback (NFB) as one of the biofeedback modalities has a very wide range of applications. Surprisingly, despite of its popularity, research on its effectiveness in many cases remains inconclusive. What is more, there are studies that have even brought contradictory results. For instance, the need to use individualised *vs* standard neurofeedback protocol is still under debate. In this article we point out the problem of the neurofeedback effectiveness underestimation which might result from the neglected neurofeedback learning inability phenomenon (also called as BCI-illiteracy). We suggest that there are three preconditions of the neurofeedback loop establishment, and subsequently, we reflect on their potential obstacles. We conclude by encouraging neurofeedback researchers and practitioners to pay more attention to observing and reporting the problem of the neurofeedback learning inability, as it is crucial factor for determining its real effectiveness.

1 Introduction

Neurofeedback is a method of creating a feedback loop between a subject's mental state and an external device. This can be done by registration of a subject's electrophysiological parameters in order to use them in an algorithm that 'translates' these parameters into an external visible feedback. There are diverse forms of such a feedback, e.g. a computer animation - in case of biofeedback training - or a control over a movement of a specific device, as in case of brain-computer interface (BCI) application [1, 2]. The main difference between the NFB training/therapy and the BCI application lays in the purpose of creating this kind of biological feedback loop. The NFB training or NFB therapy is used to help people in acquiring a desired psychophysical state, e.g. relaxation or deeper concentration. The BCI, on the other hand, in most cases serves as functional substitution of missing or paralysed limbs or other body parts. Nevertheless, the starting point of using both methods is the human ability to learn self-regulation of a given electrophysiological parameter. In this article we concentrate mostly on issues related to NFB training and therapy. However, we also refer to the BCI research results, since they might help in elucidation of some aspects of the NFB (in)efficacy.

W. P. Hunek and S. Paszkiel (Eds.): BCI 2018, AISC 720, pp. 45–58, 2018.
https://doi.org/10.1007/978-3-319-75025-5_6

Neurofeedback is a narrower term than biofeedback, as it concerns physiological correlates of the central nervous system (CNS) activity, exclusively (Fig. 1). Other physiological parameters that might be used in a biofeedback training are: e.g. temperature, skin conductance, heart-rate variability (HRV) or respiration. They have all the same functional principle (creating an algorithm that uses the registered physiological parameters and translates them into a comprehensible form of a computer feedback), but may differ in terms of complexity and field of use. Apart from electrophysiological, there are also other parameters of the CNS activity, that might be used in NFB training/therapy (Fig. 2; see also: [3, 4]).

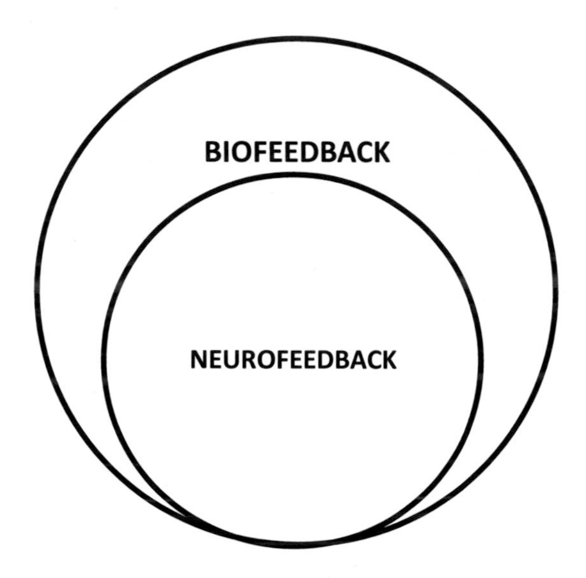

Fig. 1. The place of neurofeedback in biofeedback (Source: Szewczyk [3])

Specific functional principle of the EEG-biofeedback consists of learning the self-regulation of the electrophysiological signal generated by the brain cells. The signal is recorded by the electrodes attached to the subject's scalp and transferred through an amplifier to a computer. In order to establish this kind of neurofeedback, the electroencephalogram (EEG), which is a very complex signal, must be decomposed into several band-passes. This is done thanks to the fast Fourier transform (FFT) which differentiates the following brain waves frequencies: delta (0.5–3.5 Hz), theta (4–8 Hz), alpha (8–12 Hz), beta and gamma (>13 Hz) [5]. In a healthy human brain, each frequency attains an individual range of amplitudes. Once the signal is decomposed and the amplitude of each frequency is measured, the brain waves with their amplitudes can be depicted on a computer screen so that the subject sees how is his mental state represented in the neural activity. This is the first step to learning the biological feedback loop. If the subject knows how does his brain activity look like depending on his mental state, he is instructed by a neurotherapist (or a trainer) to increase or decrease the amplitude of a given brainwave. Since each of the brainwave frequencies is related to the human cognitive and emotional functions or - more generally - to human psychological states, it

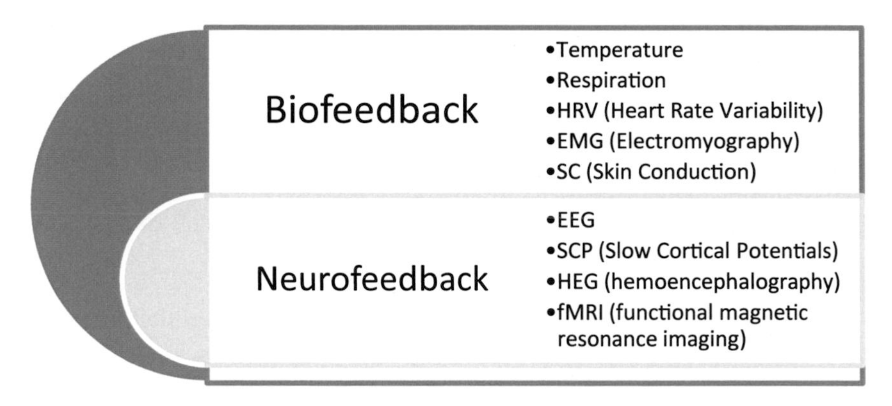

Fig. 2. Biofeedback and neurofeedback modalities (Source: Szewczyk [3])

is assumed that a change in the brainwave amplitude results in a change in the subject's mental state. The stage of gaining an intentional control over the brainwaves' characteristics is the most difficult stage of establishing the neurofeedback loop [6]. This is why the neurofeedback training is based on the rules of instrumental conditioning, consisting of symbolic visual or auditory reinforcements given to subject each time he successfully attaints the desired value of the trained parameter (for an examplary description of the threshold adjustment see: [7]). Apart from amplitudes, the neurofeedback training may also concern the interhemispheric coherence, power, z-score or other neurophysiological metrics of brain activity see: [6]. It should be emphasized that changes in the EEG pattern are evoked by the subject, exclusively, and there is no possibility of 'putting' any electrical current into the brain.

2 Area of Bio- and Neurofeedback Use and Evaluation of Its Effectiveness

Although neurofeedback was initially used to reduce the number of epileptic seizures [8], for more recent results see: [9–12], it shortly became a very popular, non-pharmacological and non-invasive method of treating many other disorders of different nature, e.g. ADHD [13–16], depression [17, 18], substance abuse [19], schizophrenia [20–22], stroke [23, 24], autistic spectrum disorder (ASD) [25, 26], emotional disorders [27–29] and tinnitus [30, 31]. Interestingly, it can be also used to enhance cognitive performance of healthy individuals [32, 33], or to enhance their memory [33, 34], peripheral visual performance [35] and even creativity and artistic performance [36–39].

The effectiveness of neuro- and biofeedback has been a widely-debated issue since the very beginnings of the neurofeedback. Yucha and Gilbert [40] presented a 5-point scale of the biofeedback effectiveness' assessment. The first level stands for Not Empirically Supported, 2nd for Possibly Efficacious, 3rd for Probably Efficacious, 4th for Efficacious and 5th for Efficacious and Specific - which means that: "The investigational treatment has been shown to be statistically superior to credible sham therapy, pill, or alternative *bona fide* treatment in at least two independent research settings." Therefore, the fact that a given disease or disorder gets one or two points does not mean that the BFB is ineffective. It might simply mean that there is still not enough evidence in order to make a reliable statement about its effectiveness. However, if the research on NFB effectiveness in a given disorder or disease is being conducted for many years, it can already be a meaningful sign of the NFB weakness.

The biofeedback efficacy was noticed in many different disorders (Table 1). Although, with reference to the question of the neurofeedback effectiveness, the hitherto studies do not allow for any conclusive statement. Many promising publications may be found but they should always be interpreted cautiously, as some methodological limitations can be observed. By the example of the ADHD, we want to show how challenging it is to determine the exact level of neurofeedback effectiveness.

Table 1. Biofeedback efficacy – state of the art in 2004 and in 2008, prepared based on Yucha and Gilbert [40]; Yucha and Montgomery [41].

Efficacy level	Diseases/disorders classified in 2004	Diseases/disorders classified in 2008
5	Incontinence (in women)	Incontinence (in women)
4	Anxiety	Anxiety
	ADHD	ADHD
	Headaches (adults)	Chronic pain
	Hypertension	Epilepsy
	Bruxism	Constipation
	Incontinence (in men)	Headaches (adults)
		Hypertension
		Motion sickness
		Raynaud's disease
		Bruxism
3	Addiction to alcohol and psychoactive substances	Addiction to alcohol and psychoactive substances
	Arthritis	Arthritis
	Chronic pain	Diabetes
	Epilepsy	Excretion disorders
	Excretion disorders	Children's headaches
	Children's headaches	Insomnia
	Insomnia	Concussion
	Concussion	Incontinence (in men)
	Vulvar Vestibulitis (vaginal pain)	Vulvar Vestibulitis (vaginal pain)

One of the first studies on ADHD treatment showed that 20 sessions of NFB may cause comparable improvements in attention and concentration to taking Ritalin [42]. Subsequent studies confirmed these results stating that the NFB may be considered as effective as pharmacotherapy in terms of reducing ADHD symptoms [43, 44]. Later research, which were conducted with more scientific rigour, i.e. randomized, double-blind placebo controlled, yielded similar positive results [45]. Even more interestingly, improvements were found to be stable on 6-months follow-up [46, 47]. This kind of evidence raised high hopes among the NFB practitioners, which is not surprising, given the side effects related to the traditional pharmacotherapy. Neurofeedback, in turn, as long as is conducted under the supervision of appropriately educated and experienced therapist, is considered as safe and having no unintended side effects [48–52]. Consistent with the aforementioned outcomes, Arns et al. [13] based on their meta-analysis, rated the efficacy of NFB in ADHD treatment as having the 5th level of Yucha and Gilbert classification.

Nevertheless, not all of the studies yield similar results. For instance, in a research by Micoulaud-Franchi et al. [53] who took into consideration assessment provided by the patients' parents, only the inattention symptoms were found to be diminished after the NFB treatment. No improvements in impulsiveness nor hyperactivity were observed. This type of evidence is somehow striking, as it is more common to overrate the NFB effectiveness based on subjective ratings, since the placebo effect has its strong influence. On the other hand, however, while the placebo effect might be the reason why the first studies seemed to be so promising, an opposite to the placebo effect has been also reported. Namely, research by Lansbergen et al. [54] showed an unexpected significant decrease of ADHD symptoms (rated by an investigator who was unaware of the subjects' group assignment) despite very low expectations of participants' from experimental as well as from control group. In fact, only 2 out of 8 subjects from the experimental group and 3 out of 6 subjects from the placebo group believed that they had been receiving a real NFB training. What is even more surprising is the fact that there were no between groups differences in terms of ADHD symptoms reduction, meaning that children who received 30 sessions of a real NFB did not outperform children who underwent 30 session of sham protocol (which was not intended to evoke any reliable improvements). Interestingly, the authors put forward some arguments in attempt to explain the lack of evidence for the NFB efficacy, which are the opposite of what other researchers claim when trying to explain similar lack of proof for NFB effectiveness, but with use of a different training protocol. Namely, Lansbergen et al. [54] suggest that the use of individualized EEG-neurofeedback (based on the results of the QEEG assessment) may be not as efficient as the standardised neurofeedback protocol, while other authors emphasize the need of using individualised protocol as potentially safer and more effective [55]. Importantly, one of the most recent meta-analysis by Cortese et al. [56] concludes that 'evidence from well-controlled trials with probably blinded outcomes does not support neurofeedback as an effective treatment for ADHD, in terms of either ADHD symptoms or other cognitive correlates' (p. 453).

The above mentioned results indicate that there is still a need for further research in order to confirm the NFB effectiveness in treating ADHD symptoms. In this context it is worth to cite some recommendations proposed by Cortese et al. [56] which should be implemented in order to ensure more valid and reliable outcomes of any future research in this topic. These are the following: 'identifying the most appropriate electrophysiological treatment target; increasing the use of standard EEG and learning protocols; developing new methods to optimize the chances that neurofeedback leads to learning at the brain level; and identifying predictors of treatment response for individual patients or at least in distinctive subgroups of children' (p. 453). In the third part of this article, we will discuss the current state of the art about the possible causes of the NFB learning inability, which has to be distinguished from the neurofeedback ineffectiveness.

3 Preconditions and Moderators of Neurofeedback Learning Ability

As it was already mentioned in the introduction, in order to establish a reliable neurofeedback loop, the three basic conditions must be fulfilled. First, the brainwaves recorded from the scalp surface reflect in the real-time the subjects' current state of mind [5, 57]. Second, once the EEG signal is decomposed and translated into a comprehensible visual or auditory (or even multisensory) feedback, the subject is able to intentionally change a given parameter of the neural activity. Third, changes in the trained parameter, evoked by the subject, should result in changes of his mental state, and these changes should occur adequately to the functional characteristics of the brainwaves' parameters. If not all of these requirements are met, any effects of the neurofeedback training could be expected. While the first assumption is already a well confirmed fact constituting the background for many currently conducted research in the field of cognitive neuroscience, the second and the third assumptions can not be taken for granted. The reason for this uncertainty is the BCI-illiteracy phenomenon, which is the inability in taking control over one's neural activity. Noteworthy are the remarks by Gruzelier [58] and Egner and Gruzelier [36] who claim that the learning ability is a crucial mediator of the neurofeedback training outcome. The same idea is put forward by Kouijzer et al. [25] and Hanslmayr et al. [59] who showed that only in these subjects who were able to learn the self-regulation of the brain activity a significant behavioural improvements were observed. In the neurofeedback literature these persons are defined as 'performers' or 'non-performers' e.g. [60], good and poor performers [61], 'regulators' and 'non-regulators' [6]. All of these terms simply mean that despite the subject's good will and his efforts put into creating the neurofeedback loop, such a person is unable to intentionally evoke the desired direction of changes in his brain activity pattern [25, 59–66]. This phenomenon is not limited to the EEG-biofeedback but may also occur in case of fMRI NFB modality [67]. The BCI-illiteracy phenomenon is as old as the research in that field itself, as Kamiya was the first to report in 1958 that 20% of his patients did not succeed in self-regulation of the trained brainwaves' amplitudes [68]. However, in spite of its long history, the problem is still present and far from being resolved. There are several potential factors that make this situation so persistent, i.e. (1) since most of the researchers simply do not report how many participants cannot establish the neurofeedback loop, there is a lack of systematic data about the extent of the BCI-illiteracy phenomenon; (2) diverse nomenclature of the phenomenon and (3) diverse methods of assessment the ability to self-regulation the neural activity which do not allow for a systematic and comprehensive literature review.

Just to give an example of the first factor, we searched for 'EEG biofeedback' and 'neurofeedback' terms in the Google Scholar search engine and get 1800 records.

Twenty-one out of 70 randomly chosen articles were considered in a deeper review (we excluded general reviews, articles on technical issues, and unobtainable papers). The authors of only two out of 21 analysed articles took into account the problem of EEG-inefficacy in their research. Although the exact percentage is very difficult to obtain, some researchers estimate that the BCI-illiteracy may be the case of up to 50% of the neurofeedback research participants [7, 59], but according to other authors' estimations this number is not higher than 25% [66] or even 21.4% [65].

With regard to the second and third factor, the authors, who mention the problem of non-performance, use many different parameters of this phenomenon, e.g. difference between the first and the last session of NFB training [69], between the baseline and the last session [65], between the resting state measured before and after the NFB training [59] or between the resting state before NFB and the change evoked by the training [59]. There are also cases of reporting more than one parameter, e.g. [7, 60, 66, 70]. For instance, Enriquez-Geppert [66] tracked how was the trained parameter changing in the course of one session as well as its changes across all of the sessions. Wan et al. [7], in turn, assessed the performance rate based on three indicators: the changes between two periods, within a short period and across the whole training time. What makes the comparisons so challenging is the fact that they should be made for each type of neurofeedback separately. The observation that a person is not able to establish a stable neurofeedback loop of one type (e.g. EEG) does not have to be equivalent to her inability to establish the neurofeedback loop with another type (e.g. fMRI). Wan et al. [7] claim that there is no universal assessment of NFB learning ability, as it should always be chosen with reference to the objective of the NFB training. For example, if the NFB training aims at improving behavioural symptoms, any neurophysiological change indicator should be correlated with the assessment of the observed behavioural changes.

The NFB learning inability problem is very often obscure, and thus neglected, by the NFB practitioners. Therapists tend to suspect their patients for the lack of engagement into the neurofeedback training, while it is not necessarily the case or not the best way to succeed. In contrast, it might be that too much effort put into the training brings the opposite results. For instance, Witte et al. [71] demonstrated that subjects' strong beliefs about their ability to control an external device may be in fact an obstacle, especially in the relaxation training. People may try to do their best in an effortful manner in order to obtain the desired amplitude value and, as a consequence, they become more focused and tensed, which results in decrease instead of increase of the SMR waves. Other type of obstacles leading to the NFB learning inability may be of biophysical nature. In some people, changes in the target feature of the electrophysiological signal are so small that it is very hard to extract them from the background noise. For instance, the P300 deflection is unobserved in 10% of the research participants [72, 73]. It does not have to indicate any neurological disorder or dysfunction, but may simply mean that the source of a given signal generation lays very deeply in a cortical sulcus so that the signal is not strong enough to reach the electrode [74]. Another example of a similar nature is the case when the target signal is 'drowned' by a strong concurrent signal generated by an adjacent cluster of neurons.

Alkoby et al. [6] in their review draw the attention to the role of specific mental strategies that may prove to be effective or not, depending on the type of the neurofeedback protocol. On one hand, Kober et al. [75] reported that in the SMR training people who used no mental strategies (i.e. did not think of anything specific) performed better in terms of the trained SMR amplitude than people who were trying to evoke some particular thoughts. On the other hand, in another protocol aimed at increasing the alpha amplitude, it is better to create some positive thoughts, e.g. picturing one's relatives or recalling other comforting memories, since these strategies help to acquire an optimal state of relaxation [34]. Nonetheless, there are also protocols for which no effect of mental strategies was found, e.g. Kober et al. [75] in case of gamma amplitude training or Hardman et al. [76] for interhemispheric coherence asymmetry training of Slow Cortical Potentials.

There is also some evidence that some psychological traits, states and abilities may moderate the NFB learning ability. Namely, individuals of higher working memory capacity and better attentional resources may be more successful in establishing the neurofeedback loop [77–79]. Nijboer et al. [80] highlights the role of the subjects' mood and motivation to perform the training. Mathiak et al. [81] showed also that people who received a kind of 'social feedback' exhibited better training outcomes in comparison to people who get a classical abstract form of feedback. Although these questions are rarely discussed in scientific papers, NFB practitioner are well aware of how important these issues are. It is important to mention the suggestion of Alkoby et al. [6] concerning the sensibility to rewards as another potential moderator of the NFB learning effectiveness. The authors note that it might be possible that the establishment of the neurofeedback loop would be easier for people who are more sensible to the external reinforcements in comparison to those who are less reward-sensitive. Their hypothesis is based on reinforcement sensitivity theory (RST) [82, 83] and on observations by Mathiak et al. [81].

Before passing to the conclusions, some issues of the NFB effectiveness that are important from the applied neurofeedback perspective will be discussed. Namely, due to the negligence of some technical aspects, even the first assumption of the NFB effectiveness is violated. It occurs, for instance, when the electrophysiological signal is drowned by the muscular artifacts resulting from jaw clenching. This problem seems to be encountered by the NFB practitioners who work with children. It is difficult to understand for some young patients what does it mean 'to be focused' or 'to be relaxed' as these are rather abstract expressions. During the first NFB sessions, children sometimes clench their jaws as a result of putting great mental and/or physical effort to increase or decrease the target brainwave amplitude. It can be relatively easily eliminated by a massage applied to this part of face or by instruction to decontract the facial muscles. It is also the therapist's task to give a clear instruction of what to do, or rather of what not to do, in order to reduce the artifacts' proportion in the recorded EEG signal. The above mentioned problem is relatively trivial, but there can be also an alternative cause of the applied neurofeedback inefficacy. While the scientific research paradigm imposes very strict rules of appropriate variables measurement and manipulation, practitioners may lack such a rigour when conducting the NFB training or therapy. Thus, many unintentional mistakes - mostly of technical nature - are possible to occur. During our practice, we saw few cases of inappropriate electrode placement, which used to be stick to the patient's hair, not even touching the scalp. We hope that continuous therapists' education and supervision will eliminate this type of erroneous practice.

4 Conclusions

Neurofeedback, as one the biofeedback's modalities is a widely used non-pharmacological and non-invasive method of treating disorders and dysfunctions of different kinds. Although its effectiveness has been investigated for many years, in many cases the research results are still inconclusive. The NFB learning inability is a potential cause of the NFB effectiveness underestimation. It is hard to determine the scale of this phenomenon because (1) very few researchers report this problem, (2) there is no consensus on the problem's terminology and (3) different methods of the NFB learning ability are used. Despite of, or rather because of these difficulties, further research should aim to elucidate the scale and the nature of the NFB learning inability, as it is crucial for a reliable NFB effectiveness estimation. Additionally, more efforts should be put in order to raise the awareness of the problem amongst NFB practitioners.

References

1. Wolpaw, J.R., Birbaumer, N., Heetderks, W.J., McFarland, D.J., Peckham, P.H., Schalk, G., Donchin, E., Quatrano, L.A., Robinson, C.J., Vaughan, T.M.: Brain-computer interface technology: a review of the first international meeting. IEEE Trans. Rehab. Eng. **8**(2), 164–173 (2000). https://doi.org/10.1109/TRE.2000.847807
2. Evans, J.R., Abarbanel, A.: Introduction to Quantitative EEG and Neurofeedback. Academic Press, Orlando (1999)
3. Szewczyk, R.: Biofeedback. In: Borkowski, P. (ed.) Biofeedback Innowacje. Akademia im. Jana Długosza w Częstochowie (2015)
4. Demos, J.N.: Getting Started with Neurofeedback. WW Norton Co., New York (2005). https://doi.org/10.1016/j.jpsychores.2005.08.007
5. Teplan, M.: Fundamentals of EEG measurement. Measur. Sci. Rev. **2**(2), 1–11 (2002). https://doi.org/10.1021/pr070350l
6. Alkoby, O., Abu-Rmileh, A., Shriki, O., Todder, D.: Can we predict who will respond to neurofeedback? A review of the inefficacy problem and existing predictors for successful EEG neurofeedback learning. Neuroscience (2017). https://doi.org/10.1016/j.neuroscience.2016.12.050
7. Wan, F., Nan, W., Vai, M.I., Rosa, A.: Resting alpha activity predicts learning ability in alpha neurofeedback. Front. Human Neurosci. **8**, 500 (2014). https://doi.org/10.3389/fnhum.2014.00500
8. Sterman, M.B., Friar, L.: Suppression of seizures in an epileptic following sensorimotor EEG feedback training. Electroencephalogr. Clin. Neurophysiol. **33**(1), 89–95 (1972). https://doi.org/10.1016/0013-4694(72)90028-4
9. Kotchoubey, B., Strehl, U., Uhlmann, C., Holzapfel, S., König, M., Fröscher, W., Blankenhorn, V., Birbaumer, N.: Modification of slow cortical potentials in patients with refractory epilepsy: a controlled outcome study. Epilepsia **42**(3), 406–416 (2001)
10. Sterman, M.B., Egner, T.: Foundation and practice of neurofeedback for the treatment of epilepsy. Appl. Psychophysiol. Biofeedback **31**(1), 21–35 (2006). https://doi.org/10.1007/s10484-006-9002-x
11. Tan, G., Thornby, J., Hammond, D.C., Strehl, U., Canady, B., Arnemann, K., Kaiser, D.A.: Meta-analysis of EEG biofeedback in treating epilepsy. Clin. EEG Neurosci. **40**(3), 173–179 (2009). https://doi.org/10.1177/155005940904000310

12. Strehl, U., Birkle, S.M., Wörz, S., Kotchoubey, B.: Sustained reduction of seizures in patients with intractable epilepsy after self-regulation training of slow cortical potentials - 10 years after. Front. Hum. Neurosci. **8**(604), 1–7 (2014). https://doi.org/10.3389/fnhum.2014.00604
13. Arns, M., de Ridder, S., Strehl, U., Breteler, M., Coenen, A.: Efficacy of neurofeedback treatment in ADHD: the effects on inattention, impulsivity and hyperactivity: a meta-analysis. Clin. EEG Neurosci. **40**(3), 180–189 (2009). https://doi.org/10.1177/155005940904000311
14. Duric, N.S., Assmus, J., Gundersen, D., Elgen, I.B.: Neurofeedback for the treatment of children and adolescents with ADHD: a randomized and controlled clinical trial using parental reports. BMC Psychiatry **12**(1), 107 (2012). https://doi.org/10.1186/1471-244X-12-107
15. Moriyama, T.S., Polanczyk, G., Caye, A., Banaschewski, T., Brandeis, D., Rohde, L.A.: Evidence-based information on the clinical use of neurofeedback for ADHD. Neurotherapeutics **9**(3), 588–598 (2012). https://doi.org/10.1007/s13311-012-0136-7
16. Sonuga-Barke, E.J.S., Brandeis, D., Cortese, S., Daley, D., Ferrin, M., Holtmann, M., et al.: Nonpharmacological interventions for ADHD: systematic review and meta-analyses of randomized controlled trials of dietary and psychological treatments. Am. J. Psychiatry **170** (3), 275–289 (2013). https://doi.org/10.1176/appi.ajp.2012.12070991
17. Choi, S.W., Chi, S.E., Chung, S.Y., Kim, J.W., Ahn, C.Y., Kim, H.T.: Is alpha wave neurofeedback effective with randomized clinical trials in depression? A pilot study. Neuropsychobiology **63**(1), 43–51 (2010). https://doi.org/10.1159/000322290
18. Dias, Á.M., van Deusen, A.: A new neurofeedback protocol for depression. Spanish J. Psychol. **14**(1), 374–384 (2011). https://doi.org/10.5209/rev
19. Sokhadze, T.M., Cannon, R.L., Trudeau, D.L.: EEG biofeedback as a treatment for substance use disorders: review, rating of efficacy, and recommendations for further research. Appl. Psychophysiol. Biofeedback **33**(1), 1–28 (2008). https://doi.org/10.1007/s10484-007-9047-5
20. Bolea, A.S.: Neurofeedback treatment of chronic inpatient Schizophrenia. J. Neurotherapy **14**(1), 47–54 (2010). https://doi.org/10.1080/10874200903543971
21. Nan, W., Chang, L., Rodrigues, J. P., Wan, F., Mak, P. U., Mak, P. I., Vai, M., Rosa, A.: Neurofeedback for the treatment of schizophrenia: case study. In: Proceedings of IEEE International Conference on Virtual Environments, Human-Computer Interfaces, and Measurement Systems, VECIMS, vol. 139, pp. 78–81 (2012). https://doi.org/10.1109/VECIMS.2012.6273182
22. Surmeli, T., Ertem, A., Eralp, E., Kos, I.H.: Schizophrenia and the efficacy of qEEG-guided neurofeedback treatment: a clinical case series. Clin. EEG Neurosci. (official journal of the EEG and Clinical Neuroscience Society (ENCS)) **43**(2), 133–144 (2012). https://doi.org/10.1177/1550059411429531
23. Doppelmayr, M., Nosko, H., Fink, A.: An attempt to increase cognitive performance after stroke with neurofeedback. Biofeedback **35**(4), 126–130 (2007). http://www.ncbi.nlm.nih.gov/pubmed/22081825
24. Rayegani, S.M., Raeissadat, S.A., Sedighipour, L., Rezazadeh, I.M., Bahrami, M.H., Eliaspour, D., Khosrawi, S.: Effect of neurofeedback and electromyographic-biofeedback therapy on improving hand function in stroke patients. Top. Stroke Rehab. **21**(2), 137–151 (2014). https://doi.org/10.1310/tsr2102-137
25. Kouijzer, M.E.J., Schie, H.T., Gerrits, B.J.L., Buitelaar, J.K., Moor, J.M.H.: Is EEG-biofeedback an effective treatment in autism spectrum disorders? A randomized controlled trial. Appl. Psychophysiol. Biofeedback **38**, 17–28 (2012). https://doi.org/10.1007/s10484-012-9204-3

26. Thompson, M., Thompson, L.: Asperger's syndrome intervention: combining neurofeedback, biofeedback and metacognition. Introduction to quantitative EEG and neurofeedback, pp. 365–415. Academic Press, San Diego (2009)
27. Otiimer, S., Othmer, S.F.: Post traumatic stress disorder. Biofeedback **37**(1), 24–31 (2009)
28. Raymond, J., Sajid, I., Parkinson, L.A., Gruzelier, J.H.: Biofeedback and dance performance: a preliminary investigation. Appl. Psychophysiol. Biofeedback **30**(1), 65–73 (2005). https://doi.org/10.1007/s10484-005-2175-x
29. Reiter, K., Andersen, S.B., Carlsson, J.: Neurofeedback treatment and Posttraumatic stress disorder. J. Nerv. Mental Dis. **204**(2), 69–77 (2016). https://doi.org/10.1097/NMD.0000000000000418
30. Schenk, S., Lamm, K., Ladwig, K.H.: Effects of a neurofeedback-based alpha training on chronic tinnitus. Verhaltenstherapie **13**(2), 2003 (2003)
31. Hartmann, T., Lorenz, I., Müller, N., Langguth, B., Weisz, N.: The effects of neurofeedback on oscillatory processes related to tinnitus. Brain Topogr. **27**(1), 149–157 (2014). https://doi.org/10.1007/s10548-013-0295-9
32. Vernon, D., Egner, T., Cooper, N., Compton, T., Neilands, C., Sheri, A., Gruzelier, J.: The effect of training distinct neurofeedback protocols on aspects of cognitive performance. Int. J. Psychophysiol. **47**(1), 75–85 (2003). https://doi.org/10.1016/S0167-8760(02)00091-0
33. Wang, J.R., Hsieh, S.: Neurofeedback training improves attention and working memory performance. Clin. Neurophysiol. **124**(12), 2406–2420 (2013). https://doi.org/10.1016/j.clinph.2013.05.020
34. Nan, W., Rodrigues, J.P., Ma, J., Qu, X., Wan, F., Mak, P.I., Mak, P.U., Vai, M.I., Rosa, A.: Individual alpha neurofeedback training effect on short term memory. Int. J. Psychophysiol. **86**(1), 83–87 (2012). https://doi.org/10.1016/j.ijpsycho.2012.07.182
35. Nan, W., Wan, F., Lou, C.I., Vai, M.I., Rosa, A.: Peripheral visual performance enhancement by neurofeedback training. Appl. Psychophysiol. Biofeedback **38**(4), 285–291 (2013). https://doi.org/10.1007/s10484-013-9233-6
36. Egner, T., Gruzelier, J.: Ecological validity of neurofeedback: modulation of slow wave EEG enhances musical performance. NeuroReport **14**(9), 1221–1224 (2003). https://doi.org/10.1097/01.wnr.0000081875.45938.d1
37. Gruzelier, J.: A theory of alpha/theta neurofeedback, creative performance enhancement, long distance functional connectivity and psychological integration. Cogn. Process. **10** (Suppl. 1), S101–S109 (2009). https://doi.org/10.1007/s10339-008-0248-5
38. Gruzelier, J.H.: EEG-neurofeedback for optimising performance. II: creativity, the performing arts and ecological validity. Neurosci. Biobehav. Rev. **44**, 142–158 (2014). https://doi.org/10.1016/j.neubiorev.2013.11.004
39. Gruzelier, J., Inoue, A., Smart, R., Steed, A., Steffert, T.: Acting performance and flow state enhanced with sensory-motor rhythm neurofeedback comparing ecologically valid immersive VR and training screen scenarios. Neurosci. Lett. **480**(2), 112–116 (2010). https://doi.org/10.1016/j.neulet.2010.06.019
40. Yucha, C., Gilbert, C.: Evidence-Based Practice in Biofeedback and Neurofeedback. Association for Applied Psychophysiology and Biofeedback (AAPB), Wheat Ridge (2004)
41. Yucha, C.B., Montgomery, D.: Evidence-based practice in biofeedback and neurofeedback. Nursing **6656** (2008). https://doi.org/10.1017/cbo9781107415324.004
42. Rossiter, T., La Vaque, T.: A comparison of EEG biofeedback and Psychostimulants in treating attention deficit/hyperactivity disorders. J. Neurotherapy **1**(1), 48–59 (1995). https://doi.org/10.1300/J184v01n01_07

43. Fuchs, T., Birbaumer, N., Lutzenberger, W., Gruzelier, J.H., Kaiser, J.: Neurofeedback treatment for attention-deficit/hyperactivity disorder in children: a comparison with Methylphenidate. Appl. Psychophysiol. Biofeedback **28**(1), 1–12 (2003). https://doi.org/10.1023/A:1022353731579
44. Rossiter, T.: The effectiveness of neurofeedback and stimulant drugs in treating AD/HD: part I. Review of methodological issues. Appl. Psychophysiol. Biofeedback **29**(2), 95–112 (2004). https://doi.org/10.1023/B:APBI.0000026636.13180.b6
45. deBeus, R.J., Kaiser, D.A.: Neurofeedback with children with attention deficit hyperactivity disorder: a randomized double-blind placebo-controlled study. In: Neurofeedback and Neuromodulation Techniques and Applications, pp. 127–152 (2011). https://doi.org/10.1016/B978-0-12-382235-2.00005-6
46. Strehl, U., Leins, U., Goth, G., Klinger, C., Hinterberger, T., Birbaumer, N.: Self-regulation of slow cortical potentials: a new treatment for children with attention-deficit/hyperactivity disorder. Pediatrics **118**(5), e1530–e1540 (2006). https://doi.org/10.1542/peds.2005-2478
47. Gevensleben, H., Holl, B., Albrecht, B., Schlamp, D., Kratz, O., Studer, P., Rothenberger, A., Moll, G.H., Heinrich, H.: Neurofeedback training in children with ADHD: 6-month follow-up of a randomised controlled trial. Eur. Child Adolesc. Psychiatry **19**(9), 715–724 (2010). https://doi.org/10.1007/s00787-010-0109-5
48. Van Dongen-Boomsma, M., Vollebregt, M.A., Slaats-Willemse, D., Buitelaar, J.K.: A randomized placebo-controlled trial of electroencephalographic (EEG) neurofeedback in children with attention-deficit/hyperactivity disorder. J. Clin. Psychiatry **74**(8), 821–827 (2013). https://doi.org/10.4088/JCP.12m08321
49. Evans, J., Rubi, M.: Ours is to reason why and how. In: Handbook of Neurofeedback: Dynamics and Clinical Applications, pp. 61–81. The Haworth Medical Press/The Haworth Press, Binghamton (2007). https://doi.org/10.1080/13533303100013927З
50. Hammond, D.C.: What is neurofeedback: an update. J. Neurotherapy **15**(4), 305–336 (2011). https://doi.org/10.1080/10874208.2011.623090
51. Holtmann, M., Stadler, C., Leins, U., Strehl, U., Birbaumer, N., Poustka, F.: Neurofeedback for the treatment of attention-deficit/hyperactivity disorder (ADHD) in childhood and adolescence. Zeitschrift Für Kinder- Und Jugendpsychiatrie Und Psychotherapie **32**(3), 187–200 (2004)
52. Little, K.D., Lubar, J.F., Cannon, R.: Neurofeedback: research-based treatment for ADHD. In: Handbook of Integrative Clinical Psychology, Psychiatry, and Behavioral Medicine: Perspectives, Practices, and Research, pp. 807–821 (2010)
53. Micoulaud-Franchi, J.-A., Geoffroy, P.A., Fond, G., Lopez, R., Bioulac, S., Philip, P.: EEG neurofeedback treatments in children with ADHD: an updated meta-analysis of randomized controlled trials. Front. Hum. Neurosci. **8**, 1–7 (2014). https://doi.org/10.3389/fnhum.2014.00906
54. Lansbergen, M.M., Van Dongen-Boomsma, M., Buitelaar, J.K., Slaats-Willemse, D.: ADHD and EEG-neurofeedback: a double-blind randomized placebo-controlled feasibility study. J. Neural Transm. **118**(2), 275–284 (2011). https://doi.org/10.1007/s00702-010-0524-2
55. Hammond, D.C.: What is neurofeedback? J. Neurotherapy (Investigations in Neuromodulation, Neurofeedback and Applied Neuroscience) **10**(4), 25–36 (2007). https://doi.org/10.1300/J184v10n04
56. Cortese, S., Ferrin, M., Brandeis, D., Holtmann, M., Aggensteiner, P., Daley, D., et al.: Neurofeedback for attention-deficit/hyperactivity disorder: meta-analysis of clinical and neuropsychological outcomes from randomized controlled trials. J. Am. Acad. Child Adolesc. Psychiatry **55**(6), 444–455 (2016). https://doi.org/10.1016/j.jaac.2016.03.007
57. Kamiya, J.: Operant control of the EEG alpha rhythm and some of its reported effects on consciousness. In: Tart, C. (ed.) Altered States of Consciousness, pp. 489–501 (1969)

58. Gruzelier, J.H.: EEG-neurofeedback for optimising performance. I: a review of cognitive and affective outcome in healthy participants. Neurosci. Biobehav. Rev. **44**, 124–141 (2014). https://doi.org/10.1016/j.neubiorev.2013.09.015
59. Hanslmayr, S., Sauseng, P., Doppelmayr, M., Schabus, M., Klimesch, W.: Increasing individual upper alpha power by neurofeedback improves cognitive performance in human subjects. Appl. Psychophysiol. Biofeedback **30**(1), 1–10 (2005). https://doi.org/10.1007/s10484-005-2169-8
60. Weber, E., Köberl, A., Frank, S., Doppelmayr, M.: Predicting successful learning of SMR neurofeedback in healthy participants: methodological considerations. Appl. Psychophysiol. Biofeedback **36**(1), 37–45 (2011). https://doi.org/10.1007/s10484-010-9142-x
61. Doehnert, M., Brandeis, D., Straub, M., Steinhausen, H.-C., Drechsler, R.: Slow cortical potential neurofeedback in attention deficit hyperactivity disorder: is there neurophysiological evidence for specific effects? J. Neural Transm. **115**(10), 1445–1456 (2008). https://doi.org/10.1007/s00702-008-0104-x
62. Kotchoubey, B., Strehl, U., Holzapfel, S., Blankenhorn, V., Fröscher, W., Birbaumer, N.: Negative potential shifts and the prediction of the outcome of neurofeedback therapy in epilepsy. Clin. Neurophysiol. **110**(4), 683–686 (1999). https://doi.org/10.1016/S1388-2457(99)00005-X
63. Kropotov, J.D., Grin-Yatsenko, V.A., Ponomarev, V.A., Chutko, L.S., Yakovenko, E.A., Nikishena, I.S.: ERPs correlates of EEG relative beta training in ADHD children. Int. J. Psychophysiol. **55**(1), 23–34 (2005). https://doi.org/10.1016/j.ijpsycho.2004.05.011
64. Escolano, C., Aguilar, M., Minguez, J.: EEG-based upper alpha neurofeedback training improves working memory performance. In: Proceedings of the Annual International Conference of the IEEE Engineering in Medicine and Biology Society, EMBS, pp. 2327–2330 (2011). https://doi.org/10.1109/IEMBS.2011.6090651
65. Zoefel, B., Huster, R.J., Herrmann, C.S.: Neurofeedback training of the upper alpha frequency band in EEG improves cognitive performance. NeuroImage **54**(2), 1427–1431 (2011). https://doi.org/10.1016/j.neuroimage.2010.08.078
66. Enriquez-Geppert, S., Huster, R.J., Scharfenort, R., Mokom, Z.N., Zimmermann, J., Herrmann, C.S.: Modulation of frontal-midline theta by neurofeedback. Biol. Psychol. **95**(1), 59–69 (2013). https://doi.org/10.1016/j.biopsycho.2013.02.019
67. Scheinost, D., Stoica, T., Wasylink, S., Gruner, P., Saksa, J., Pittenger, C., Hampson, M.: Resting state functional connectivity predicts neurofeedback response. Front. Behav. Neurosci. **8**, 338 (2014). https://doi.org/10.3389/fnbeh.2014.00338
68. Kamiya, J.: The first communications about operant conditioning of the EEG. J. Neurotherapy **15**(1), 65–73 (2011). https://doi.org/10.1080/10874208.2011.545764
69. Dekker, M.K.J., Sitskoorn, M.M., Denissen, A.J.M., Van Boxtel, G.J.M.: The time-course of alpha neurofeedback training effects in healthy participants. Biol. Psychol. **95**(1), 70–73 (2014). https://doi.org/10.1016/j.biopsycho.2013.11.014
70. Enriquez-Geppert, S., Huster, R.J., Herrmann, C.S.: Boosting brain functions: improving executive functions with behavioral training, neurostimulation, and neurofeedback. Int. J. Psychophysiol. **88**(1), 1–16 (2013). https://doi.org/10.1016/j.ijpsycho.2013.02.001
71. Witte, M., Kober, S.E., Ninaus, M., Neuper, C., Wood, G.: Control beliefs can predict the ability to up-regulate sensorimotor rhythm during neurofeedback training. Front. Hum. Neurosci. **7**, 8 (2013). https://doi.org/10.3389/fnhum.2013.00478
72. Polich, J.: Normal variation of P300 from auditory stimuli. Electroencephalogr. Clin. Neurophysiol./Evoked Potentials **65**(3), 236–240 (1986). https://doi.org/10.1016/0168-5597(86)90059-6

73. Conroy, M.A., Polich, J.: Normative variation of P3a and P3b from a large sample: gender, topography, and response time. J. Psychophysiol. **21**(1), 22–32 (2007). https://doi.org/10.1027/0269-8803.21.1.22
74. Allison, B.Z., Neuper, C.: Could anyone use a BCI? In: Tan, D., Nijholt, A. (eds.) Brain-Computer Interfaces, pp. 35–54 (2010). https://doi.org/10.1007/978-1-84996-272-8
75. Kober, S.E., Witte, M., Ninaus, M., Neuper, C., Wood, G.: Learning to modulate one's own brain activity: the effect of spontaneous mental strategies. Fron. Hum. Neurosci. **7**, 1–12 (2013). https://doi.org/10.3389/fnhum.2013.00695
76. Hardman, E., Gruzelier, J., Cheesman, K., Jones, C., Liddiard, D., Schleichert, H., Birbaumer, N.: Frontal interhemispheric asymmetry: self regulation and individual differences in humans. Neurosci. Lett. **221**(2–3), 117–120 (1997). https://doi.org/10.1016/S0304-3940(96)13303-6
77. Roberts, L.E., Birbaumer, N., Rockstroh, B., Lutzenberger, W., Elbert, T.: Self-report during feedback regulation of slow cortical potentials. Psychophysiology **26**(4), 392–403 (1989). https://doi.org/10.1111/j.1469-8986.1989.tb01941.x
78. Daum, I., Rockstroh, B., Birbaumer, N., Elbert, T., Canavan, A., Lutzenberger, W.: Behavioural treatment of slow cortical potentials in intractable epilepsy: neuropsychological predictors of outcome. J. Neurol. Neurosurg. Psychiatry **56**(1), 94–97 (1993). https://doi.org/10.1136/jnnp.56.1.94
79. Wangler, S., Gevensleben, H., Albrecht, B., Studer, P., Rothenberger, A., Moll, G.H., Heinrich, H.: Neurofeedback in children with ADHD: specific event-related potential findings of a randomized controlled trial. Clin. Neurophysiol. (official journal of the International Federation of Clinical Neurophysiology) **122**(5), 942–950 (2011). https://doi.org/10.1016/j.clinph.2010.06.036
80. Nijboer, F., Furdea, A., Gunst, I., Mellinger, J., McFarland, D.J., Birbaumer, N., Kübler, A.: An auditory brain-computer interface (BCI). J. Neurosci. Meth. **167**(1), 43–50 (2008). https://doi.org/10.1016/j.jneumeth.2007.02.009
81. Mathiak, K.A., Alawi, E.M., Koush, Y., Dyck, M., Cordes, J.S., Gaber, T.J., et al.: Social reward improves the voluntary control over localized brain activity in fMRI-based neurofeedback training. Front. Behav. Neurosci. **9**, 136 (2015). https://doi.org/10.3389/fnbeh.2015.00136
82. Gray, J.A.: The neuropsychology of temperament. In: Strelau. J. (ed.) Explorations in Temperament: International Perspectives on Theory and Measurement, pp. 105–128. Plenum Press, London (1991). https://doi.org/10.1007/978-1-4899-0643-4_8
83. Gray, J.A.: Framework for a taxonomy of psychiatric disorder. In: Van Goozen, S.H.M., van de Poll, N.E., Sergeant, J.A. (eds.) Emotions: Essays on Emotion Theory, pp. 29–59. Lawrance Erlbaum, Hove (1994)

Influence of LED Radiation on Brain Electrical Activity Record

Zolubak Magda[1(✉)] and Mariusz Pelc[1,2]

[1] Opole University of Technology,
ul. Proszkowska 76, 45-278 Opole, Poland
{m.zolubak,m.pelc}@po.opole.pl
[2] CIS Department, University of Greenwich,
Park Row, London SE10 9LS, UK
m.pelc@greenwich.ac.uk

Abstract. This paper deals with influence of LED radiation on brain wave signals. Diodes that emit light radiation for room lighting are being increasingly used, and are also used in tablet screens, laptops, televisions, and smartphones. Recent research shows that the light emitted by the LEDs affects various physiological functions because they emit light in the 400–490 nm range, which is blue. This work presents the effect of the LED light coming from the computer monitor on the brain electrical activity record.

Keywords: Neurofeedback · EEG activity monitoring

1 Introduction

Without a doubt the light has a huge influence on the way a human being is functioning not only in the physiological but also mental meaning. Because of that we can consider two different ways this influence may be considered:

- Therapeutic effect may be used as means of medical treatment against certain mental health issues,
- Pathogenic effect (resulting from too much exposure related for example a specific job that requires using computers, monitors, etc.) may ultimately lead to violating a very delicate chemical and electrical brain balance and hence have a very negative impact on an individual's mental condition.

Numerous research reveal that light deficit may affect mood, especially when in some seasons, periodically, there is significantly less light. It is very likely related to de-regulation of the circadin rhythm. For example, animals kept in dark light, roughly 50 lux, were showing symptoms of depressive behaviour [5]. There is more and more new therapies that are using various sorts of light, including muscle therapy, brain-related disorders, etc. TCLT (Transcranial LED Therapy) is a very effective therapy using LED light for trating disorders that

W. P. Hunek and S. Paszkiel (Eds.): BCI 2018, AISC 720, pp. 59–68, 2018.
https://doi.org/10.1007/978-3-319-75025-5_7

require improvements of the brain blood circulation decreasing among others cell apoptosis [6,7].

More and more often attention is paid to harmfulness of the LED light in the context of blue light emission, which according to the latest research may damage eye receptors. This kind of light is emitted by bulbs, monitors, etc. In general, the LED light colour is very important as certain results reveal that blue LED light of higher colour correlation may lead to eye lense epithelium [4,10].

2 Related Work

Both, the light intensity and the source of light may have a huge impact on various biological processes, including circadian rhythm, and this is typical not only for animals, like mice [1], but also human beings.

Various tests conducted on people by the use of electroluminescent diode (LED) have proven that this kind of light does not only propagate through sighting (which is perfectly understandable), but also they permeate through skull leading to regulating of certain cognitive functions but also improving pathological states of brain [2]. Additionally, it was determined that to much exposure to a LED light may affect many physiological functions and hence it may be used therapeutically against circadian rhythm disorders as well as sleeping-related issues.

Apart from the source of light, what may alternate the light influence is also its colour. For example, too much exposure to blue light may lead to photoreceptors damages as well as retina damages [3,9]. It has also been investigated that LED light, depending on its colour may increase attention and relaxation levels or work quite the opposite way [8].

In this work the main focus is on monitoring changes occurring in human brain once exposed to LED light. Analysis of frequency spectrum reveals changes affecting different types of brain waves.

3 Test Group and Research Methodology

Only 5 adult people were subject of the designed tests to determine influence of the LED light on the electric brain activity record - 2 men and 3 women, all aged 25–35 years old. All of them had very high qualifications (MSc degree) and were spending similar amount of time in front of the LED monitors. None of the persons takes medication of any kind that would potentially affect the EEG activity. The low number of people involved into all tests was intentional as the aim of our research was to not to provide some statistically meaningful analysis but rater do some kind of preliminary study of the problem and depending on its results carry on with a bigger scale research (subject to future work).

Using QEEG method all the subjects were thoroughly tested with regard to determine particular influence of the LED light on their brain waves characteristics after 90 min exposure to this type of radiation. The research was carried out in so called *mid line* area, which means C3, C4, Cz and Fz places (see Fig. 1).

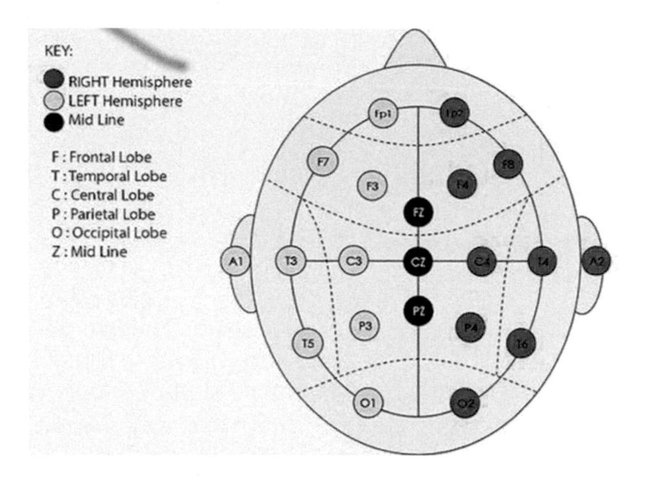

Fig. 1. Positions of the QEEG electrodes

The brain waves frequency ranges are gathered in the Table 1.

Table 1. Brain waves frequency ranges

Delta	Theta	Alpha	SMR	Beta1	Beta2
0.5–4.0 Hz	4.0–8.0 Hz	8.0–12.0 Hz	12.0–15.0 Hz	15.0–22.0 Hz	22.0–55.0 Hz

Based on the EEG signals recorded with these persons in the aforementioned places, the average amplitudes for the selected waves frequency ranges (absolute values) and their percentage share in the whole spectrum (relative values) as well as average values for amplitudes for the whole spectrum in these places. Then all the registered signals were filtered and a number of spectrograms were obtained.

As a reference sample the EEG signals acquired from people who had not used any LED emitting device since at least 8 h. The signals recording time was very similar for all tested people. Then all the people were asked to watch continuously for at least 90 min some not very much emotionally engaging content on the internet, such as mems, films and pictures. In general, this content was intended to be considered nice or neutral the least. After 90 min of watching and being exposed to the LED light from monitor those people were tested again. Each test, including the reference sample and the sample after LED light exposure, took 2 min in the sequence: 1 min with open eyes (simulating day-time brain activity) and 1 min with closed eyes (simulating night-time/sleep brain activity). None of these tests was of diagnostic purpose but only to register changes that appear in the EEG record after being exposed to the LED light.

3.1 Filtration Method

In order to focus on the relevant frequency range we used a cascade filter (rather than band-pass filter) comprising of:

- low-pass filter with the cut off frequency of 49 Hz,
- notch filter (to eliminate the power supply network frequency components) with the cut off frequency 50 Hz,
- high-pass filter with the cut-off frequency of 1 Hz.

The filters themselves were not very much sophisticated as we decided to stick with the classical digital filters. After carrying out a number of numerical experiments we decided to choose the Chebyshev II filter mainly due to its satisfactory frequency selectivity. All tests and filtrations were performed in Matlab using a number of dedicated script files.

The main filtration was being done in the *processEEG.m* file of the structure shown in the Fig. 2.

```
% EEG frequency analysis. Script to import EDF data in Matlab
% Clearing whole Matlab workspace and closing all figures
clear all;
close all;
clc;
% Load EDF data files
[hdr, alldata] = LoadEdfFile('Person1Data.edf');
% Select one out of 4 registered data channels
ch1 = alldata(1,1:4096);
% Example analysis for channel 1
[ch1] = FilterNotch(ch1);
[ch1] = FilterLowPass(ch1);
[ch1] = FilterHighPass(ch1);
% Display Example Spectogram
[s,f,t,p]=spectrogram(ch1,hanning(512),511,[1:80],250,'yaxis');
figure;
spectrogram(ch1,hanning(512),511,[1:80],250,'yaxis');
ylim([0 30])
```

Fig. 2. Main Matlab script performing the whole frequency analysis

The *LoadEdfFile()* function is another Matlab script of form (see Fig. 3):

```
function [hdr, alldata] = LoadEdfFile(fname)
% Prompt the user to choose the file
[hdr, alldata] = edfread(fname);
end
```

Fig. 3. The *LoadEdfFile.m* script to read EDF data

The *edfread()* function seen in the *LoadEdfFile.m* script is one of available Matlab functions.

```
function [eegdata_f] = FilterNotch(eegdata_f)
% Notch filter
N_ch=size(eegdata_f,1);
fsamp = 250;         % sampling frequency
nf_f_low = 49;       % low-pass frequency
nf_f_high = 51;      % high-pass frequency
order = 2;           % 2nd order
RippleF = 80;        % ripple factor
disp(N_ch);
% Chebyshev II filter used
[bn,an]=cheby2(order,RippleF,[nf_f_low nf_f_high]/(fsamp/2),'stop');
% Applying the filter to the EEG data
for i=1:N_ch
    eegdata_f(i,:)=filtfilt(bn,an,eegdata_f(i,:));
end
end
```

Fig. 4. The *FilterNotch.m* script to implement notch filter

The filters used for the filtration purposes (notch, low-pass and high-pass) are shown in Figs. 4, 5 and 6. The notch filter is set to eliminate 50 Hz harmonic present in the power supply.

The low-pass filter is set to cut-off frequencies above 60 Hz, but in practice, due to non-ideal filter characteristic the cut-off band is rather closer to 50 Hz, which is fine for the considered brain waves frequency ranges.

The cut-off frequency set to 1 Hz in practice (due to non-ideal filter characteristic) results in bandwidth starting around 2 Hz. As a result all filters produce "working" range between around 2–55 Hz.

4 Results and Interpretations

As mentioned before, all the tests were carried out with 5 adults. However, in order to illustrate how the LED light affects brain waves we have selected most representative results for 4 people.

The biggest change in the EEG signal in comparison to the reference sample were spotted in the 5–15 Hz range, which includes from the point of view of bio-feedback EEG analysis Theta waves (4–8 Hz), Alpha waves (8–12 Hz) and SMR rhythm (12–15 Hz). This maps to EEG Alpha waves (8–13 Hz) and Beta (14–30 Hz). In the Fig. 7 one can see filtration results for the Fz point for 4 out of 5 people.

In the Fig. 8 one can see changes that occurred in the same place after 90 min spent in front of the LCD monitor. There is easy to notice significantly less area of dark colour (lower amplitudes of given frequencies).

```
function [eegdata_f] = FilterLowPass(eegdata_f)
N_ch=size(eegdata_f,1);
fsamp = 250;        % sampling frequency
lp_f_low = 60;      % cut-off frequency
order = 5;          % 5th order
RippleF = 80;       % ripple factor
disp(N_ch);
% Low-pass EEG filter
[b,a]=cheby2(order,RippleF,lp_f_low/(fsamp/2),'low');
for i=1:N_ch
    eegdata_f(i,:)=filtfilt(b,a,eegdata_f(i,:));
end
end
```

Fig. 5. The *FilterLowPass.m* script to cut-off high frequencies

```
function [eegdata_f] = FilterHighPass(eegdata_f)
N_ch=size(eegdata_f,1);
fsamp = 250;        % sampling frequency
hp_f_high = 1;      % cut-off frequency
order = 5;          % 5th order
RippleF = 80;       % ripple factor
disp(N_ch);
% High-pass EEG filter
[b,a]=cheby2(order,RippleF,hp_f_high/(fsamp/2),'high');
for i=1:N_ch
    eegdata_f(i,:)=filtfilt(b,a,eegdata_f(i,:));
end
end
```

Fig. 6. The *FilterHighPass.m* script to cut-off low frequencies

As one can see from the Fig. 8, 3 out of 5 people have noticeably increased values of Alpha waves for Fz, C3 and C43 points (see additionally in Fig. 9). In case of Beta1 and SMR rhythm, if amplitudes were increasing, it was for all persons whilst if amplitudes were decreasing, it also was for all persons.

There was only one exception, for one person (the fourth subject) the SMR rhythm decreased in all measured points and Beta1 increased (see additionally in Fig. 11).

In case of the slow Theta waves, for 2 people it has increased in the temple area, for the remaining people it has decreased in all measured points (see Fig. 10).

It is also noticeable that the Theta waves frequency changes depending on whether a given person has open or closed eyes (see Fig. 12).

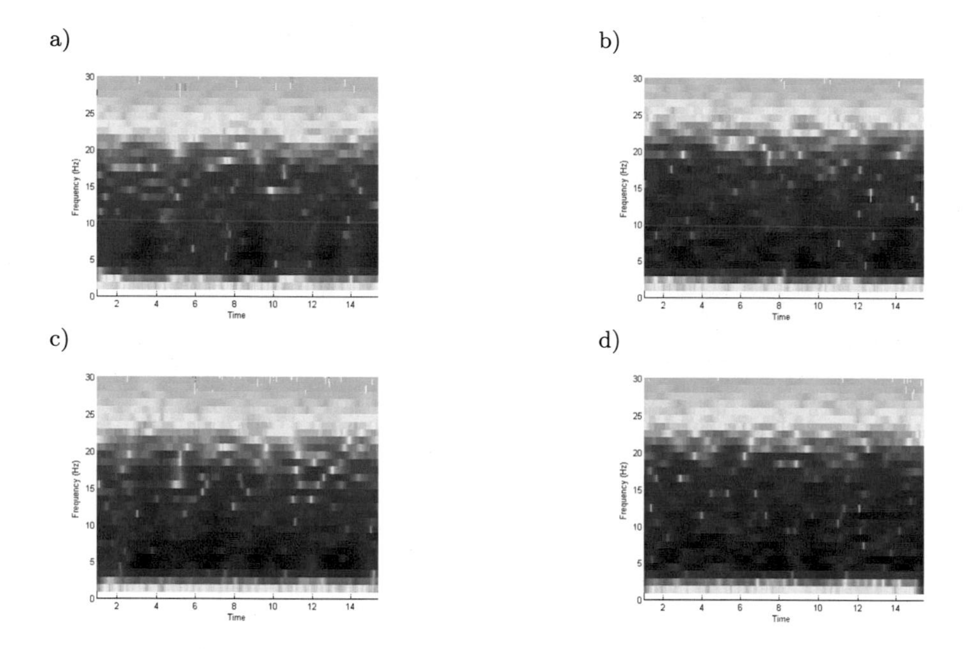

Fig. 7. Reference sample for the Fz place for 4 out of 5 testet people

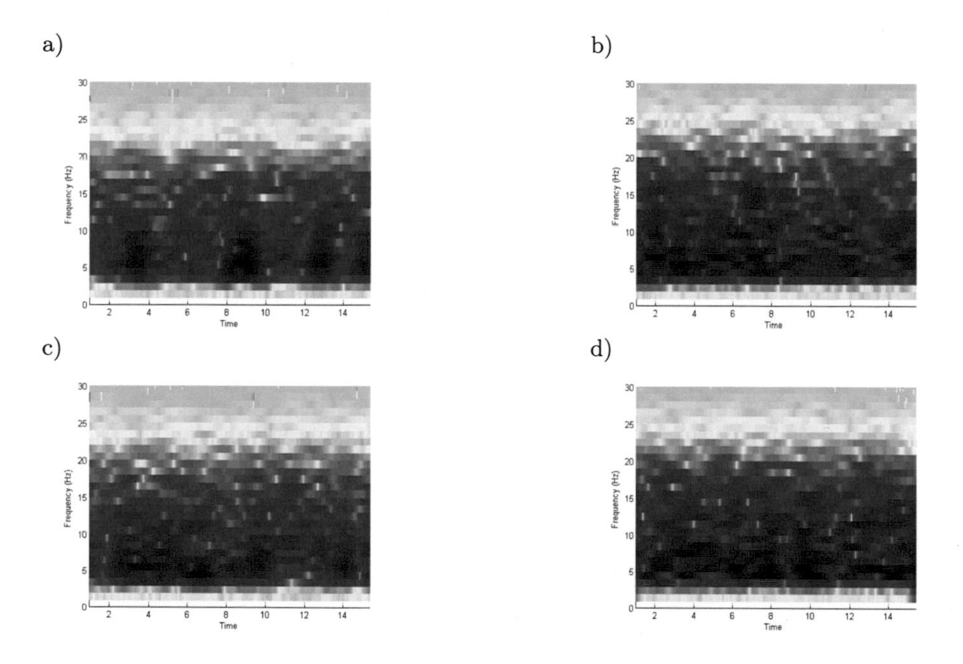

Fig. 8. Results for the the Fz place for 4 out of 5 testet people after 90 min exposure to the LED light

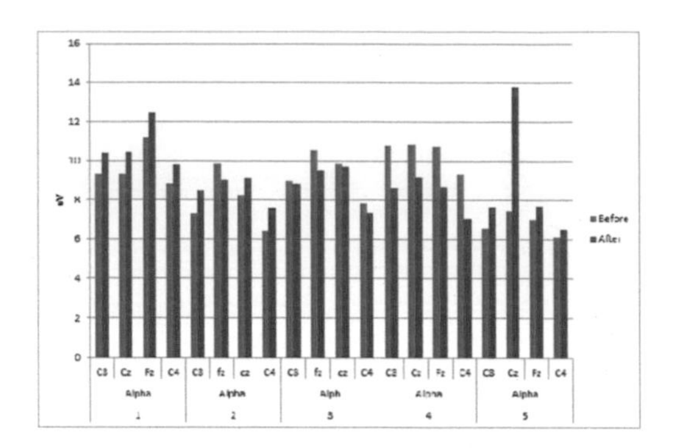

Fig. 9. Alpha waves frequency change for the tested people

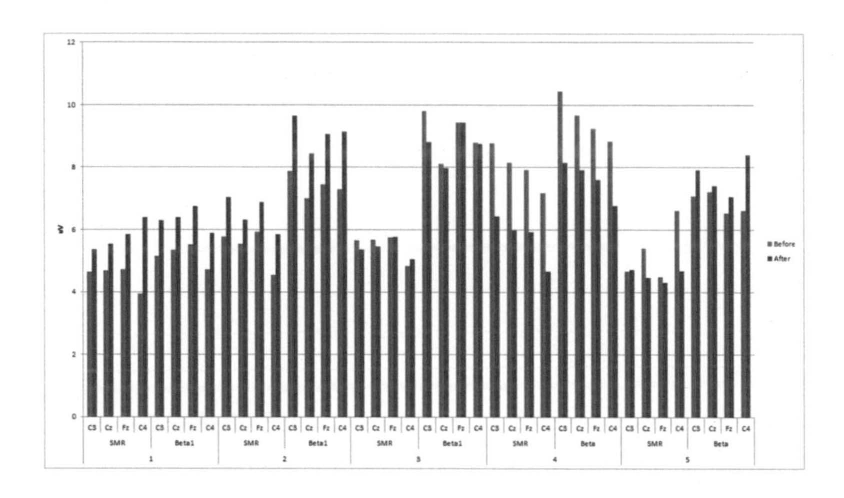

Fig. 10. SMR rhythm and Beta1 frequency change for the tested people

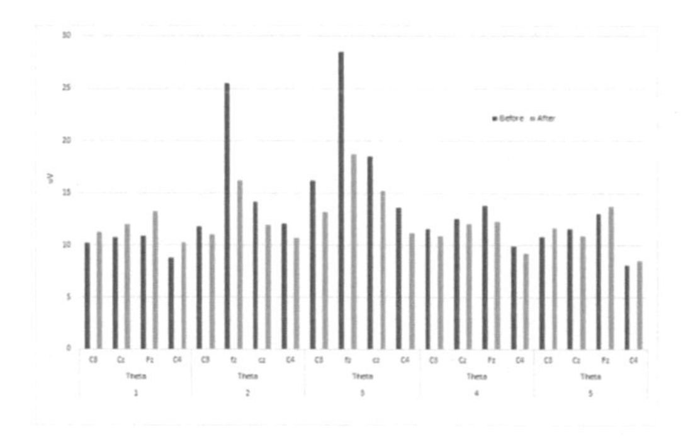

Fig. 11. Theta frequency change for the tested people

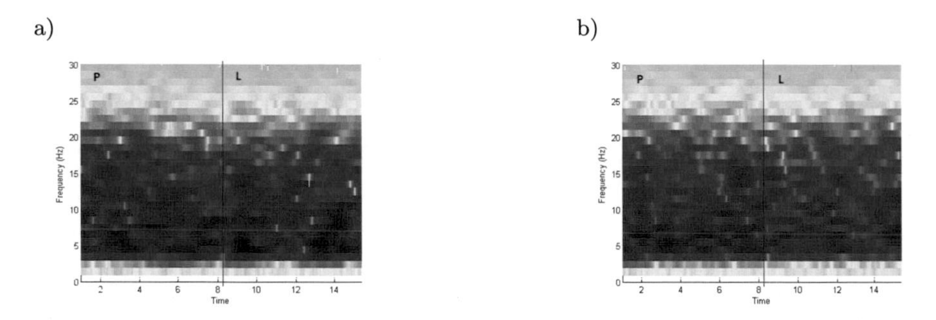

Fig. 12. Visible drop of the Theta for closed eyes (L) after LED light exposure; (a) reference sample, (b) after exposure to the LED light

5 Conclusions

Although the research we conducted was of preliminary nature, certain observations are definitely worth commenting on and may be drive for further research involving more representative sample of people.

First of all, changes for the open eyes within Theta wave frequency range may suggest that the LSED light may be of significance with regard to sleep regulation because it can affect positively the falling asleep phase through increasing the slow waves but in the same time increase in the slow waves frequency range may negatively affect sleep quality. In the spectrograms it is easy to spot that for tested people when they had their eyes closed the Theta waves frequency visibly decreased in comparison to the reference sample. This unambiguously would confirm that the LED light affects sleep regulation (as shown in Fig. 12 for one of the tested people).

Increase of the Alpha waves frequency for open eyes for tested people, as long as it remains within norm, may positively affect passive attention and memory. After 90 min exposure to the LED light one can see visible and measurable changes which after such a relatively short exposure do not affect in great extent on the brain functioning. This is because this does not concern all measurable frequencies in full range but only some bigger changes within 5–15 Hz frequency range. But if the exposure time would increase, this may lead to affecting the circadian rhythm.

References

1. Alves-Simoes, M., Coleman, G., Canal, M.M.: Effects of type of light on mouse circadian behaviour and stress levels. Lab. Anim. **50**(1), 21–29 (2015)
2. Moghadam, H.S., Nazari, M.A., Jahan, A., Mahmoudi, J., Salimi, M.M.: Beneficial effects of transcranial Light Emitting Diode (LED) therapy on attentional performance: an experimental design. Iran. Red Crescent Med. J. 8 (2017)
3. Jaadane, I., Boulenguez, P., Chahory, S., Carre, S., Savoldelli, M., Jonet, L., Behar-Cohen, F., Martinsons, C., Torriglia, A.: Retinal damage induced by commercial Light Emitting Diodes (LEDs). Free Radic. Biol. Med. **84**, 373–384 (2015)

4. Kuse, Y., Ogawa, K., Tsuruma, K., Shimazawa, M., Hara, H.: Damage of photoreceptor-derived cells in culture induced by light emitting diode-derived blue light. Sci. Rep. **4**, 5223 (2014)
5. Leach, G., Adidharma, W., Yan, L.: Depression-like responses induced by daytime light deficiency in the diurnal grass rat (Arvicanthis niloticus). PLoS ONE **8**(2), e57115 (2013)
6. Vani, A.A., Miranda, E.F., de Carvahlo, P.D.T.C., Dal Corso, S., Bjordal, J.M., Leal-Junior, E.C.P.: Effect of phototherapy (LLLT and LEDT) on exercise Performance and markers of exercise recovery: a systematic review with meta-analysis. Physioterapy **101**(Supplement 1), e1580–e1581 (2015)
7. Salgado, S.I.A., Parreira, R.B., Ceci, L.A., de Oliveira, L.V.F., Zangaro, R.A.: Transcranial light emitting diode therapy (TCLT) and its effects on neurological disorders. Bioeng. Biomed. Sci. **5**(1), 1 (2015)
8. Shin, J.-Y., Chun, S.-Y., Lee, C.-S.: Analysis of the effect on attention and relaxation level by correlated color temperature and illuminance of LED lighting using EEG signal. J. Korean Inst. Illum. Electr. Install. Eng. **27**, 9–17 (2013)
9. Tosini, G., Ferguson, I., Tsubota, K.: Effects of blue light on the circadian system and eye physiology. Mol. Vis. J. **22**, 61–72 (2016)
10. Xie, C., Li, X., Tong, J., Yangshun, G.: Effects of white Light-Emitting Diode (LED) light exposure with different Correlated Color Temperatures (CCTs) on human lens epithelial cells in culture. Photochem. Photobiol. **90**(4), 853–859 (2014)

Using Neurofeedback as an Alternative for Drug Therapy in Selected Mental Disorders

Zolubak Magda[1(✉)] and Mariusz Pelc[1,2]

[1] Opole University of Technology,
ul. Proszkowska 76, 45-278 Opole, Poland
{m.zolubak,m.pelc}@po.opole.pl
[2] CIS Department, University of Greenwich,
Park Row, London SE10 9LS, UK
m.pelc@greenwich.ac.uk

Abstract. Neurofeedback is a very effective technique that may be considered as an alternative to drug therapy to support development of children with different mental disorders such as ADD. The aim of this paper is to show how neurofeedback can be used for therapeutic purposes to increase selected signal parameters. For the test purposes 30 therapeutic sessions with 13-year old boy are presented showing theta waves frequency and SMR increase in the frontal part of head and partial improvements in the central part. Another sessions lead to bringing the relevant signals frequencies near to desired, correct values. Overall outcome of the training has lead to increased ability to concentrate and to lengthening the time through which the patient was able to work without distractions.

Keywords: Neurofeedback · Mental therapy

1 Introduction

Human's brain is a very complex system and it diagnostics and treatment is extremely difficult task. Brain functioning and its overall performance is strongly dependent on a very delicate balance between its chemical and electrical activity which nature and dependencies have not yet been fully discovered (despite research going on since actually decades) but on the other side, due to a very significant level of individual features any potential any potential therapy must be strongly individualised.

The most typical method that is used to address various kind of brain dysfunctions are medication-based. A patient is being diagnosed regarding the type and level of dysfunction and an appropriate medication is being prescribed to bring the brain functioning back to balance. The problem is that there is virtually no medication that would be 100% side effects free (assuming that the therapy is carried out correctly as incorrect therapy may lead to even further

W. P. Hunek and S. Paszkiel (Eds.): BCI 2018, AISC 720, pp. 69–84, 2018.
https://doi.org/10.1007/978-3-319-75025-5_8

worsening of the patient condition). The situation becomes particularly problematic when its aim becomes humans brain. What is needed in order to achieve the desired outcome, not only the proper choice of medicine is critical but also its very specific dosing. Underdose or overdose may ultimately lead to completely unpredictable consequences (for example, it may handicap mental capabilities of the patients).

As a very promising alternative to the drug-based therapy, one can use neurofeedback methods and targeted training. An experienced trainer may run certain preliminary tests (like registering EEG waves) and spot some deviations from "norm". Then, after planning a series of trainings and using neurofeedback to assess their effects on the brain functioning the trainer may mitigate certain problems and improve overall patient's health condition. In this paper we are aiming to show this very approach in application to a teenager boy with ADD symptoms.

In the reminder of this paper we'll describe our approach in more detail, provide results, then we'll discuss progress and it's all outcomes.

2 Related Work

The QEEG signals have already proven to provide sufficient information for diagnosing various kind of disorders or brain impairments. In general, QEEG stands for Quantitative Electroencephalography and it is, as the name indicates, quantitative and not qualitative method in the contrary to EEG. As such it provides more information that EEG because it may provide information about percentage share of different kind of brain waves and their amplitudes, Fast Fourier Transform, etc.

Typically the diagnosis is being done by running a very specific tests where the patients revealing certain symptoms of some sort of abnormal mental functioning. Such tests have a repetitive nature where patients are exposed to the same or very similar stimuli, every time being very carefully monitored so that some sort of progress analysis could be performed. A series of such specialised and dedicated tests can be for example used to detect autism or classify epilepsy attacks.

Measuring QEEG signals in a specific place allows to detect abnormalities as well as to determine some typical deviations for a given disorders in selected regions of the human brain [4]. As far as autistic children are concerned, typical efficiency of verification of this kind of disorder based on signals analysis may reach around 95% [2,8]. But as such the QEEG is useful in order to detect changes characteristic for autism spectrum, and its efficiency is primarily lays in actually indicating that there is something wrong whilst some dedicated psychologically-development tests may actually fully confirm the case and also assess how advanced it is.

Monitoring the EEG signals continuously (using neurofeedback method) for patients suffering e.g. from epilepsy allows to classify epilepsy attacks and also allows quantitative analysis to create patterns for epilepsy attacks and resultingly - to predict how likely it is that the patient will have such attacks in the future. A very exact cognition of the brain neurophysiology along with widely understood neuroimaging contribute to a better and quicker diagnosis of such disorders as Attention Deficit Disorder (ADD) or Attention Deficit Hyperactivity Disorder (ADHD) [6,7]. This kind of diagnostics brings equally positive effects with elderly people [14].

Neurofeedback as such can not only be used for diagnostics purposes but also for widely understood therapy. And even if results achieved using bio-feedback cannot compete with medication, it certainly does have significantly less side effects as far as ADHD therapy is concerned [5]. Combining various neuroimaging techniques in real-time allows testing many brain regions at the same time, which is exceptionally useful among others in neurosurgery, potentially limiting the extent of resection [15]. Diagnostics results using QEEG signals suggest, that measuring signal spectrum power for alpha and beta waves can represent translative pharmaco-dynamic bio-marker for proving functional effects selected medicines [9].

As mentioned before, detecting many of the above disorders is doable on-line during recording the EEG signal. However, as far as the diagnostics based on the EEG signal is concerned, a holistic view on various deviations of the signal from norm is needed due to the fact that typically one kind of mental disorder can trigger changes of different signal parameters. Here comes the expert knowledge which allows to narrow down a very specific few markers and map them into a specific type of abnormality. This knowledge can be expressed (at least to a level) in a policy.

In [12] some results of therapy applied to 59 children with ADHD are presented and discussed. The focus of this work is to compare neurofeedback therapy (biofeedback-EEG) with other behavioural or pharmacological therapies. It was proven that the neurofeedback therapy improves attention and behavioural response deficits which could not have been achieved using the behavioural or pharmacological therapies [12]. Furthermore, neurofeedback was also tested in OCD (obsessive compulsive disorder), ASD (Autism Spectrum Disorder), GAD (Generalised Anxiety Disorder) as well as depression proving that neurofeedback was more effective than passive or semi-active treatment [1]. In the biofeedback-EEG therapy its efficiency is guaranteed by regularity and appropriate selection of protocols adjusted to individual patients needs [11]. Neurofeedback as therapy is not only applicable to children aged 7–10 years old, but also adults, as indicated in [3,11].

In this work we show results of neurofeedback therapy applied to a teenager boy as an alternative to pharmacological therapy.

3 Training and Tests

In this section we'll provide a very detailed explanation regarding what kind of problems were spotted while recording the EEG signal for the teenager boy, how the training looked liked and what kind of progress was being made throughout the process and finally, we'll show some evident improvements in certain areas of the brain activity. A very thorough discussion of the results will allow to understand the ground from which our statements and conclusions have come.

3.1 Equipment Used

For the purpose of signal analysis all the signals and data presented in this section were recorded using QEEG equipment shown in the Fig. 1.

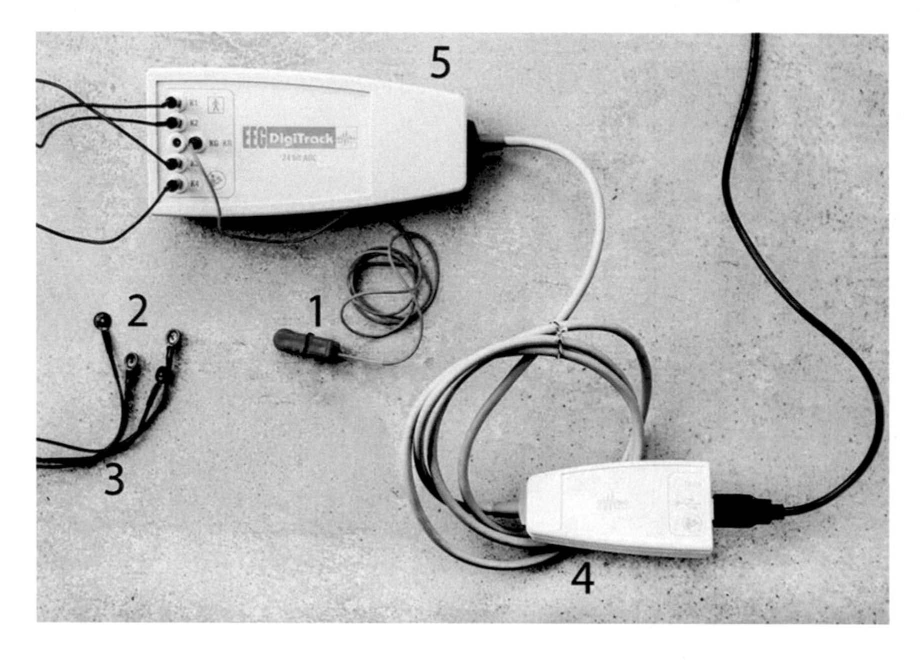

Fig. 1. QEEG equipment

In the Fig. 1 one can see annotations from 1 to 5 which identify the following system components:

1. Reference electrode.
2. Cup-shape electrode.
3. Cabling.
4. Measurement interfaces.
5. Transducer used to analyse/pre-process the signal.

During real-time EEG signal acquisition using the QEEG equipment one can monitor any undesirable changes such as seizures disorder and other abnormalities in brain functioning.

3.2 Test Scenario

In order to show how the neurofeedback-based therapy looks like, we have arranged a very simple test scenario comprising of the following steps:

- initial evaluation - in this step the QEEG equipment was used to acquire a set of different brain waves to determine a base-line for a teenager boy diagnosed with ADD,
- a set of 10 trainings was applied and the boy's condition was re-evaluated in order to determine if there is any need for further training,
- a set of 10 follow-by trainings was re-applied until the brain waves have reached the desired levels.

In terms of the trainings, they use gamification techniques where patients play an interactive game having electrodes fixed to their scalp. As it is practically impossible to fully control any particular brain wave, what usually happens is that the patient is trying to somehow change state of his mind by thinking of something, focusing on something, using relaxation techniques, etc. This obviously affects some/all brain waves. By watching effects on the screen (the way the game responds to the state of mind changes) he can train himself to reach such a state of mind on demand and as a result - modify electrical and chemical features of the brain. Which is to a level comparable to what happens when a patient is being medicated.

In order to determine EEG levels for all relevant wave types, the QEEG electrodes should be positioned precisely in appropriate and relevant brain regions. For the signal acquisition purposes the electrodes were placed in the following places: C3, C4, Fz, Cz and additionally P3, P4, F3, F4. Location of these places is shown in Fig. 2.

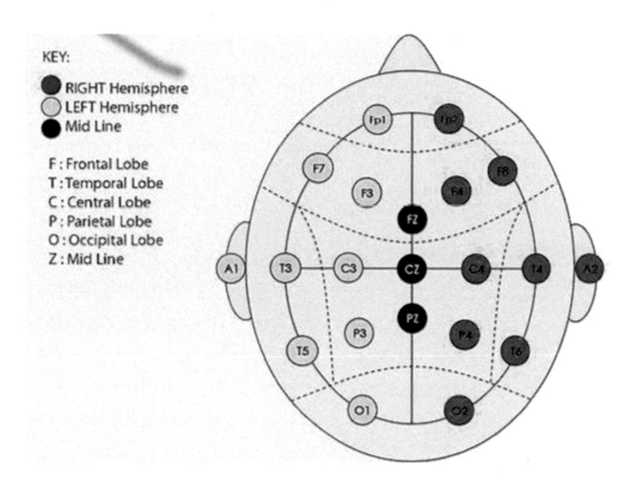

Fig. 2. Example brain waves registered with the QEEG equipment showing learning curve

Frequency ranges for the relevant brain waves are as follows:

- Delta: 0.5–4.0 Hz,
- Theta: 4.0–8.0 Hz,

- Alpha: 8.0–12.0 Hz,
- SMR: 12.0–15.0 Hz,
- Beta: 15.0–22.0 Hz,
- Beta2: 22.0–50.0 Hz.

In the Fig. 3 one can see example time characteristics for all relevant brain wave types (Delta, Theta, Alpha, Beta2, Beta1) and SMR rhythm. Some of the traces indicate some artifacts (see Delta waves). In turn, in the Beta2 time trace one can clearly see the so called "learning curve".

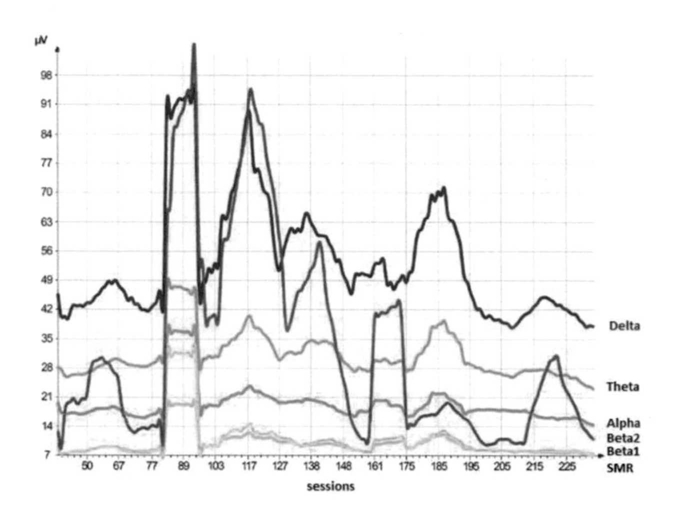

Fig. 3. Example brain waves registered with the QEEG equipment with visible learning curve

3.3 Results and Interpretations

Based on control (preliminary) EEG signals acquisition, in each of the previously mentioned points, a number of average values of all the waves types amplitudes were registered (their absolute values) and their percentage contribution to the whole spectrum (their relative values) as well as average amplitudes values for the whole spectrum in the mentioned points. The signals were acquired with the patient having first open, then closed eyes. All results were gathered in the Tables 1 and 2.

Interpretation of the baseline results. These results are treated as baseline results to which those results obtained after certain number of trainings will be compared. As one can see, although the Theta waves are within norm, their amplitude are on the high side, especially in the F3, Cz and Fz places. Such raised Theta waves amplitudes may lead to worse remembering, inability to focus, etc. Alpha wave percentage is lowered with amplitudes raised at the back of head. The SMR is significantly lowered which ultimately may lead to so called focus tension. Beta1 waves lowered especially at the mid line.

Table 1. Base line signal recording - open eyes

Wave	Left hemisphere			Mid line			Right hemisphere		
	Point	uV	%	Point	uV	%	Point	uV	%
Delta	**F3**	68.35	51.0	**Fz**	98.71	54.5	**F4**	63.91	51.6
	C3	52.21	44.2	**Cz**	58.39	44.9	**C4**	51.74	46.8
	P3	33.26	37.1	–	–	–	**P4**	47.34	44.2
Theta	**F3**	29.94	22.3	**Fz**	43.14	23.8	**F4**	27.96	22.6
	C3	25.65	21.7	**Cz**	41.14	24.0	**C4**	27.96	22.6
	P3	18.86	21.0	–	–	–	**P4**	17.82	16.6
Alpha	**F3**	14.66	10.9	**Fz**	15.94	8.80	**F4**	12.67	10.2
	C3	17.33	14.7	**Cz**	15.92	12.3	**C4**	12.23	11.1
	P3	17.06	19.0	–	–	–	**P4**	15.54	14.5
SMR	**F3**	6.32	4.7	**Fz**	7.19	4.0	**F4**	5.55	4.5
	C3	6.85	5.8	**Cz**	7.37	5.7	**C4**	6.24	5.6
	P3	7.05	7.9	–	–	–	**P4**	6.29	5.9
Beta1	**F3**	7.05	5.3	**Fz**	7.20	4.0	**F4**	6.72	5.4
	C3	7.09	6.0	**Cz**	7.23	5.9	**C4**	6.60	6.0
	P3	6.94	7.7	–	–	–	**P4**	6.72	6.3
Beta2	**F3**	7.76	5.8	**Fz**	9.10	5.0	**F4**	7.09	5.9
	C3	9.0	7.6	**Cz**	9.42	7.2	**C4**	8.97	8.1
	P3	6.46	7.2	–	–	–	**P4**	13.39	12.5

Interpretation of the baseline results. The amplitude values for Delta waves are above norm and amplitude values for Theta waves are raised in all measurement points. Similarly for the Alpha waves. The SMR rhythm amplitude values and Beta2 waves within norm, however, the percentage share is a bit lowered.

Interpretation of the results after 10 trainings. Trainings were carried out in the mid line area (meaning that here was the main training focus) in the C3 and Cz places. As a result the Alpha waves levels have dropped and the Theta waves, although still above norm, have also dropped in comparison to the baseline results. The SMR rhythm amplitudes are within norm despite slightly lowered percentage share. Beta1 waves amplitudes, despite a bit lowered percentage share, remain at appropriate level.

Interpretation of the results after 10 trainings. Closer look at the amplitude levels and percentage share clearly shows that for Alpha and Theta waves they have visibly dropped from their initial values. However, majority of them are still above norm either, with the amplitude or percentage shares respect (Tables 3, 4 and 5).

Interpretation of the results after 12 trainings. Increased level of Delta waves is very likely resulting from artifacts, like eye blinking. Percentage share for Theta waves in places a bit above norm or just below, only in Fz and Cz places

Table 2. Base line signal recording - closed eyes

	Left hemisphere			Mid line			Right hemisphere		
	Point	uV	%	Point	uV	%	Point	uV	%
Delta	**F3**	98.52	58.3	**Fz**	87.99	55.0	**F4**	76.91	56.3
	C3	52.91	39.8	**Cz**	54.17	39.6	**C4**	44.01	40.1
	P3	40.78	27.3	–	–	–	**P4**	36.13	26.0
Theta	**F3**	28.61	16.9	**Fz**	28.66	17.9	**F4**	27.67	19.3
	C3	25.68	19.3	**Cz**	31.21	22.8	**C4**	23.98	22.0
	P3	37.93	25.4	–	–	–	**P4**	35.42	25.5
Alpha	**F3**	21.75	12.9	**Fz**	21.73	13.6	**F4**	20.48	14.3
	C3	30.35	22.8	**Cz**	26.39	19.3	**C4**	20.80	19.1
	P3	44.53	29.8	–	–	–	**P4**	43.35	31.2
SMR	**F3**	6.42	3.8	**Fz**	6.09	3.8	**F4**	5.97	4.2
	C3	6.98	5.3	**Cz**	7.30	5.3	**C4**	5.89	5.4
	P3	9.22	6.2	–	–	–	**P4**	8.34	6.0
Beta1	**F3**	7.20	4.3	**Fz**	7.09	3.8	**F4**	5.97	4.2
	C3	7.48	7.8	**Cz**	7.09	4.4	**C4**	6.63	4.6
	P3	10.08	6.7	–	–	–	**P4**	8.95	6.4
Beta2	**F3**	6.61	3.9	**Fz**	8.42	5.3	**F4**	5.77	4.0
	C3	8.60	6.5	**Cz**	8.99	6.6	**C4**	7.57	7.0
	P3	6.97	4.24.7		–	–	**P4**	5.80	4.1

at high level. But the level is lower than previously. SMR rhythm and Beta2 lowered with respect to percentage share (Table 6).

Interpretation of the results after 20 trainings. Increased level of Delta and Theta may indicate tiredness as being so called "slow waves" they increase while tiredness or sleepiness. The same happens with the Alpha waves after closing eyes.

Interpretation of the results after 30 trainings. In comparison to the situation after 20 trainings, one can see that the Theta waves amplitudes have dropped the norm or very little above. Actually, the only waves that very visibly stand out and are significantly above norm are Alpha waves. Incorrect percentage share for SMR rhythm (Tables 7 and 8).

Interpretation of the results after 30 trainings. One can observe significant drop of Alpha, Delta and Theta waves in comparison to signals registered after 20 trainings. Overall, they are all slightly above norm.

Table 3. Signals acquired after 10 trainings - open eyes

Wave	Left hemisphere			Mid line			Right hemisphere		
	Point	uV	%	Point	uV	%	Point	uV	%
Delta	**F3**	48.99	41.3	**Fz**	40.47	36.7	**F4**	56.79	44.9
	C3	29.89	30.2	**Cz**	31.74	29.7	**C4**	29.12	34.4
	P3	27.70	28.3	–	–	–	**P4**	25.10	29.1
Theta	**F3**	24.99	21.1	**Fz**	28.38	25.7	**F4**	26.34	20.8
	C3	24.04	24.3	**Cz**	30.14	28.2	**C4**	22.26	26.3
	P3	23.22	23.7	–	–	–	**P4**	22.23	25.8
Alpha	**F3**	15.49	13.0	**Fz**	17.93	16.3	**F4**	13.94	11.0
	C3	22.48	22.7	**Cz**	21.59	20.2	**C4**	14.44	17.1
	P3	18.48	18.9	–	–	–	**P4**	19.11	22.2
SMR	**F3**	6.64	5.6	**Fz**	6.37	5.8	**F4**	5.30	4.2
	C3	6.66	6.7	**Cz**	6.78	6.3	**C4**	5.47	6.5
	P3	7.46	7.6	–	–	–	**P4**	5.93	6.9
Beta1	**F3**	6.83	5.8	**Fz**	7.59	6.9	**F4**	6.33	5.0
	C3	7.99	8.1	**Cz**	8.45	7.9	**C4**	6.66	7.9
	P3	7.22	7.4	–	–	–	**P4**	6.54	7.6
Beta2	**F3**	15.73	13.3	**Fz**	9.59	8.7	**F4**	17.79	14.1
	C3	8.01	8.1	**Cz**	8.27	7.7	**C4**	6.71	7.9
	P3	13.75	14.1	–	–	–	**P4**	7.34	8.5

Table 4. Signals acquired after 10 trainings - closed eyes

Wave	Left hemisphere			Mid line			Right hemisphere		
	Point	uV	%	Point	uV	%	Point	uV	%
Delta	**F3**	70.42	44.9	**Fz**	38.37	29.7	**F4**	59.23	39.5
	C3	34.72	26.8	**Cz**	36.05	26.3	**C4**	29.16	25.8
	P3	40.21	24.1	–	–	–	**P4**	34.01	24.6
Theta	**F3**	33.24	21.2	**Fz**	38.64	30.0	**F4**	36.53	24.4
	C3	37.21	28.7	**Cz**	42.62	31.1	**C4**	35.18	31.2
	P3	44.54	26.8	–	–	–	**P4**	39.54	28.6
Alpha	**F3**	22.27	14.2	**Fz**	23.65	18.3	**F4**	22.14	14.8
	C3	29.42	22.7	**Cz**	28.71	20.9	**C4**	22.32	19.8
	P3	43.46	26.1	–	–	–	**P4**	35.94	26.0
SMR	**F3**	7.38	4.7	**Fz**	9.04	7.0	**F4**	7.27	4.8
	C3	10.19	7.9	**Cz**	10.90	7.9	**C4**	9.73	8.6
	P3	12.61	7.6	–	–	–	**P4**	11.57	8.4
Beta1	**F3**	7.48	4.8	**Fz**	8.76	6.8	**F4**	6.95	4.6
	C3	9.51	7.3	**Cz**	10.01	7.3	**C4**	8.52	7.5
	P3	10.87	6.5	–	–	–	**P4**	9.42	6.8
Beta2	**F3**	16.07	10.2	**Fz**	10.55	8.2	**F4**	17.73	11.8
	C3	8.66	6.7	**Cz**	8.93	6.5	**C4**	7.97	7.1
	P3	14.58	8.8	–	–	–	**P4**	7.91	5.7

Table 5. Signals acquired after 20 trainings - open eyes

Wave	Left hemisphere			Mid line			Right hemisphere		
	Point	uV	%	Point	uV	%	Point	uV	%
Delta	**F3**	51.86	48.0	**Fz**	59.75	49.4	**F4**	41.61	44.7
	C3	38.95	36.2	**Cz**	41.94	38.1	**C4**	37.73	41.3
	P3	31.43	35.1	–	–	–	**P4**	26.56	33.7
Theta	**F3**	23.61	21.9	**Fz**	25.30	20.9	**F4**	21.81	23.4
	C3	21.96	20.4	**Cz**	25.70	23.4	**C4**	20.60	22.5
	P3	19.75	22.3	–	–	–	**P4**	18.53	23.5
Alpha	**F3**	12.20	11.3	**Fz**	16.40	13.6	**F4**	11.50	12.4
	C3	24.82	23.0	**Cz**	19.88	18.1	**C4**	17.17	16.6
	P3	15.45	17.4	–	–	–	**P4**	14.35	18.2
SMR	**F3**	5.91	5.5	**Fz**	6.41	5.3	**F4**	5.50	5.9
	C3	7.23	6.7	**Cz**	7.35	6.7	**C4**	6.06	6.6
	P3	7.84	8.8	–	–	–	**P4**	6.74	8.6
Beta1	**F3**	6.51	6.6	**Fz**	6.90	5.7	**F4**	5.84	6.3
	C3	7.78	7.2	**Cz**	7.81	7.1	**C4**	6.18	6.8
	P3	6.71	7.6	–	–	–	**P4**	5.95	7.5
Beta2	**F3**	7.93	7.3	**Fz**	6.13	5.1	**F4**	6.86	7.4
	C3	6.96	6.5	**Cz**	7.25	6.6	**C4**	5.69	6.2
	P3	7.55	8.5	–	–	–	**P4**	6.69	8.5

Table 6. Signals acquired after 20 trainings - closed eyes

Wave	Left hemisphere			Mid line			Right hemisphere		
	Point	uV	%	Point	uV	%	Point	uV	%
Delta	**F3**	60.03	44.9	**Fz**	41.91	33.2	**F4**	47.63	39.4
	C3	36.52	27.2	**Cz**	39.37	27.2	**C4**	33.18	29.2
	P3	37.75	23.3	–	–	–	**P4**	33.58	23.6
Theta	**F3**	32.74	24.5	**Fz**	36.51	28.9	**F4**	33.85	28.7
	C3	35.40	26.4	**Cz**	43.53	30.1	**C4**	32.59	28.7
	P3	48.78	30.2	–	–	–	**P4**	46.70	32.8
Alpha	**F3**	19.94	14.9	**Fz**	24.27	19.2	**F4**	19.45	16.1
	C3	34.38	25.6	**Cz**	33.29	23	**C4**	25.15	22.1
	P3	41.66	25.8	–	–	–	**P4**	33.3	23.4
SMR	**F3**	7.43	5.6	**Fz**	7.55	6.0	**F4**	7.37	6.1
	C3	9.30	6.9	**Cz**	9.49	6.6	**C4**	7.64	6.7
	P3	13.6	8.4	–	–	–	**P4**	12.26	8.6
Beta1	**F3**	7.02	5.3	**Fz**	8.91	7.1	**F4**	6.69	5.5
	C3	9.30	6.9	**Cz**	10.78	7.4	**C4**	8.69	7.6
	P3	12.11	7.5	–	–	–	**P4**	10.14	7.1
Beta2	**F3**	6.43	4.8	**Fz**	7.14	5.7	**F4**	6.05	5.0
	C3	8.11	6.0	**Cz**	8.36	5.8	**C4**	6.36	5.6
	P3	7.81	4.8	–	–	–	**P4**	6.59	8.5

Table 7. Signals registered after 30 trainings - open eyes

Wave	Left hemisphere			Mid line			Right hemisphere		
	Point	uV	%	Point	uV	%	Point	uV	%
Delta	**F3**	25.91	31.5	**Fz**	30.05	36.3	**F4**	23.77	33.6
	C3	31.26	38.5	**Cz**	27.35	30.7	**C4**	36.61	43.6
	P3	26.99	30.4	–	–	–	**P4**	23.85	29.7
Theta	**F3**	19.99	24.3	**Fz**	20.83	25.2	**F4**	17.39	24.6
	C3	18.50	22.8	**Cz**	23.18	26.0	**C4**	19.47	23.2
	P3	18.76	21.1	–	–	–	**P4**	17.38	21.6
Alpha	**F3**	17.54	21.3	**Fz**	15.10	18.3	**F4**	13.08	15.5
	C3	15.43	19.0	**Cz**	19.29	21.6	**C4**	13.05	15.5
	P3	23.74	26.7	–	–	–	**P4**	21.09	26.2
SMR	**F3**	7.04	8.6	**Fz**	5.77	7.0	**F4**	5.66	8.0
	C3	5.42	6.7	**Cz**	6.89	7.7	**C4**	4.97	5.9
	P3	7.70	8.7	–	–	–	**P4**	6.95	8.7
Beta1	**F3**	6.72	8.2	**Fz**	6.08	7.4	**F4**	6.09	8.6
	C3	5.99	7.4	**Cz**	7.18	8.1	**C4**	5.42	6.5
	P3	6.82	7.7	–	–	–	**P4**	6.26	7.8
Beta2	**F3**	4.99	6.1	**Fz**	4.84	5.9	**F4**	4.83	6.8
	C3	4.66	5.7	**Cz**	5.23	5.9	**C4**	4.42	5.3
	P3	4.85	5.5	–	–	–	**P4**	4.81	

Table 8. Signals registered after 30 trainings - closed eyes

empty	Left hemisphere			Mid line			Right hemisphere		
	Point	uV	%	Point	uV	%	Point	uV	%
Delta	**F3**	32.18	29.6	**Fz**	33.44	31.7	**F4**	27.63	28.9
	C3	28.75	30.1	**Cz**	33.86	29.1	**C4**	31.62	34.1
	P3	32.24	22.4	–	–	–	**P4**	33.04	23.2
Theta	**F3**	29.53	27.1	**Fz**	30.64	29.1	**F4**	27.76	29.1
	C3	27.00	28.3	**Cz**	33.69	28.9	**C4**	25.95	27.9
	P3	41.53	28.9	–	–	–	**P4**	40.55	28.5
Alpha	**F3**	24.57	22.6	**Fz**	21.33	20.3	**F4**	19.66	20.6
	C3	19.91	20.8	**Cz**	25.46	21.9	**C4**	17.63	19.0
	P3	42.10	29.3	–	–	–	**P4**	42.59	29.9
SMR	**F3**	8.93	8.2	**Fz**	7.49	7.1	**F4**	7.79	8.2
	C3	7.59	7.9	**Cz**	9.06	7.8	**C4**	6.58	7.1
	P3	11.10	7.7	–	–	–	**P4**	10.08	7.1
Beta1	**F3**	8.34	7.7	**Fz**	7.55	7.2	**F4**	7.88	8.3
	C3	7.48	7.8	**Cz**	8.87	7.6	**C4**	6.58	7.1
	P3	10.73	7.5	–	–	–	**P4**	10.36	7.3
Beta2	**F3**	5.30	4.9	**Fz**	4.88	4.6	**F4**	4.74	5.0
	C3	4.76	5.0	**Cz**	5.50	4.7	**C4**	4.50	4.8
	P3	6.09	4.2	–	–	–	**P4**	5.80	4.1

4 Conclusion

The aim of this paper was to show an alternative to mediciation-based therapy for people suffering from ADD. For this purpose we have carried out a number of tests where a number of patients were applied QEEG-based therapy. During this therapy the trainer was controlling all its aspects, starting from the pace/intensity up to selecting the most appropriate, individualised training.

The therapy was in a way iterative, because after each 10 of training sessions the previous results were fed into next sessions for adjustment purposes. This allowed to make actually constant improvement of the patients condition. The results we gathered during the therapy sessions clearly show that such an approach can bring measurable and visible effects. The relevant electrical brain activity (especially the Theta waves) changes in response to increasing the number of training sessions, reaching in the end close-to-norm (or even within-norm) levels. And that is being achieved with virtually not side effects whatsoever, which are so much typical for traditional, medication-based therapies.

Of course, the sample (number of people) was not statistically meaningful hence we are fully aware that drawing some far-fetched conclusions based on the obtained results might be a bit of a stretch. However, without a doubt those results are interesting enough to plan some more tests, especially including more subjects (patients).

References

1. Begemann, M.J.H., Florisse, E.J.R., van Lutterveld, R., Kooyman, M., Sommer, I.E.: Efficacy of EEG neurofeedback in psychiatry: a comprehensive overview and meta-analysis. Transl. Brain Rhythm. **1**(1), 19–29 (2016)
2. Coben, R., Mohammad-Rezazadeh, I., Cannon, R.L.: Using quantitative and analytic EEG methods in the understanding of connectivity in autism spectrum disorders: a theory of mixed over- and under-connectivity. Front. Hum. Neurosci. (2014)
3. Cowley, B., Holmström, E., Juurmaa, K., Kovarskis, L., Krause, C.M.: Computer enabled neuroplasticity treatment: a clinical trial of a novel design for neurofeedback therapy in adult ADHD. Front. Hum. Neurosci. (2016)
4. Franko, E., Wehner, T., Joly, O., Lowe, J., Porter, M., Kenny, J., Thompson, A., Rudge, P., Collinge, J., Mead, S.: Quantitative EEG parameters correlate with the progression of human prion diseases. J. Neurol. Neurosurg. Psychiatr. **87**(10), 1061–1067 (2016)
5. Geladé, K., Janssen, T., Bink, M., van Mourik, R.: Behavioral effects of neurofeedback compared to stimulants and physical activity in attention-deficit/hyperactivity disorder: a randomized controlled trial. J. Clin. Psychiatry **77**(10), e1270–e1277 (2016)
6. Heinrich, H., Strehl, U., Arns, M.: Neurofeedback and ADHD. Front. Hum. Neurosci. **9**, 602 (2016)
7. Husain, A.M., Shina, S.R.: Continuous EEG Monitoring Principles and Practice. Springer, Cham (2017)
8. Pop-Jordanova, N., Zorcec, T., Demerdzieva, A., Gucev, Z.: QEEG characteristics and spectrum weighted frequency for children diagnosed as autistic spectrum disorder. Nonlinear Biomed. Phys. **4**, 4 (2010)

9. Keavy, D., Bristow, L.J., Sivarao, D.V., Batchelder, M., King, D., Thangathirupathy, S., Macor, J.E., Weed, M.R.: The QEEG signature of selective NMDA NR2B negative allosteric modulators; a potential translational biomarker for drug development. PLoS ONE **11**(4), e0152729 (2016)
10. Kubik, A.: Training Bio-feedback for Specialist and Therapist, vol. 2. Elmico Publishing House, Poland (2015). (in Polish)
11. Mohagheghi, A., Amiri, S., Bonab, N.M., Chalabianloo, G., Noorazar, S.G., Tabatabaei, S.M., Farhang, S.: A randomized trial of comparing the efficacy of two neurofeedback protocols for treatment of clinical and cognitive symptoms of ADHD: theta suppression/beta enhancement and theta suppression/alpha enhancement. BioMed Res. Int. **2017**, 7 (2017)
12. Moreno-García, I., Meneres-Sancho, S., de Rey, C.C.-V.: A randomized controlled trial to examine the posttreatment efficacy of neurofeedback, behavior therapy, and pharmacology on ADHD measures. J. Atten. Disorders (2017)
13. Pelc, M., Anthony, R.: Towards policy-based self-configuration of embedded systems. Syst. Inf. Sci. Notes **2**(1), 20–26 (2007)
14. Reis, J., Portugal, A., Fernandes, L., Afonso, N., Pereira, M., Sousa, N., Dias, N.: An alpha and theta intensive and short neurofeedback protocol for healthy aging working-memory training. Front. Aging Neurosci. **8**, 157 (2016)
15. Skrap, M., Main, D., Ius, T., Fabbro, F., Tomasion, B.: Brain mapping: a novel intraoperative neuropsychological approach. J. Neurosurg. **125**(4), 877–887 (2016)
16. Trans Cranial Technologies: 10/20 System Positioning Manual, Hong Kong (2012)
17. Ward, P., Pelc, M., Hawthorne, J., Anthony, R.: Embedding dynamic behaviour into a self-configuring software system. In: Autonomic and Trusted Computing. LNCS, vol. 5060, pp. 373–387. Springer, Heidelberg (2008)

Statistical Methods for Analyzing Deceleration and Acceleration Capacity of the Heart Rate

Mirosław Chyliński(✉) and Mirosław Szmajda

Opole University of Technology, 45-758 Opole, Poland
m.chylinski@doktorant.po.edu.pl, m.szmajda@po.opole.pl

Abstract. This paper presents the application of statistical methods for Electrocardiography (ECG) examinations. Computer program designed to assessment of deceleration capacity (DC) and acceleration capacity (AC) of the heart rate in the LabVIEW environment with Biomedical Toolkit can be used to analyze the health status of patients with heart diseases or patients after myocardial infarction (MI). A study on ECG recordings from Holter or biomedical data files is executed using specific tools of Digital Signal Processing (DSP).

Keywords: Deceleration capacity · Acceleration capacity
LabVIEW Biomedical Toolkit · Heart rate · Electrocardiography
Holter ECG

1 Introduction

Deceleration capacity is a strong predictor of mortality for patients after myocardial infarction and is much more precise than left-ventricular ejection fraction (LVEF) [1,2]. According to the research, patients after myocardial infarction have a low DC value and their mortality is greater than the healthy subjects [1]. Healthy individuals have a DC parameter greater than 4, 5 ms [1]. An important element for the analysis of the ECG signal from Holter recordings is recognition of R-R intervals from the QRS complex and the noise and artifacts elimination.

In this article, the authors present three programs designed in LabVIEW environment and use the LabVIEW Biomedical Toolkit for ECG analysis as well as assessment of deceleration capacity and acceleration capacity of recorded signals. Computer applications used LabVIEW Advanced Signal Processing Toolkit and LabVIEW 2016 Digital Filter Design Toolkit as well. The aim of the presented study was to examine whether the age of people after myocardial infarction has an influence on DC and AC parameters. No researches have been found that might define the dependance between age and DC and AC values for heart muscle. It is assumed that both parameters are not dependent on age of people. Received results obtained to verify the null hypothesis, assuming, that the patient's age does not affect the value of the DC and AC parameters for the patient after myocardial infarction.

W. P. Hunek and S. Paszkiel (Eds.): BCI 2018, AISC 720, pp. 85–97, 2018.
https://doi.org/10.1007/978-3-319-75025-5_9

2 Study Population

The study group consisted of 26 adults with heart disease of both sex at age of 18–74 years. Recorded Twelve-lead ECG signals was downloaded from PhysioBank ATM and used St. Petersburg INCART 12-lead Arrhythmia Database (incartdb). Examination covered 13 males and 13 females. In the four subjects diagnosed coronary artery disease, arterial hypertension and left ventricular hypertrophy, in the three subjects diagnosed earlier MI, in the three subjects diagnosed transient ischemic attack, in the one subject diagnosed ventricular trigeminy, in the one subject diagnosed sinus node dysfunction and in the eight subjects recognized another heart diseases [5]. The subjects were divided into two age subgroups: 18–58 and 59–74 years.

3 Methods

3.1 Calculating of R-R Intervals

In order to perform DC and AC estimations, the author's designed in LabVIEW program: Preprocessing and Identification (PaI) recognizes R waves from the ECG signal. Front panel of Preprocessing and Identification program is shown in Fig. 1.

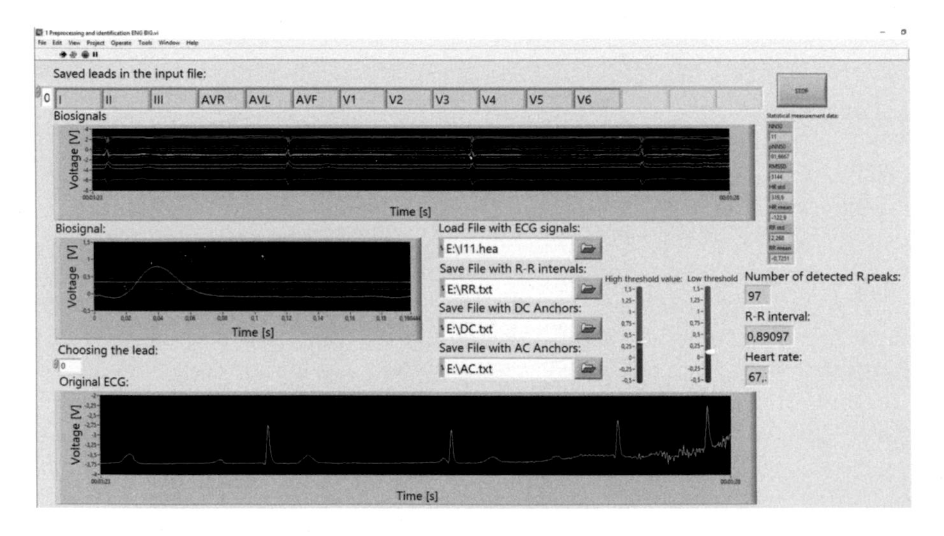

Fig. 1. View of the Front Panel of the Preprocessing and identification program in LabVIEW environment.

The following R-R intervals were counted and saved to a file. The PaI program also enumerates the anchors DC and anchors AC as well and write data to the files. It is also possible to select the appropriate ECG channel whose QRS complexes are the most expressive.

3.2 Artifacts Elimination

An author's program: Elimination of 5% (AE) for eliminating artifacts has been designed to remove those intervals R-R whose values are greater or less than 5% of the average value. Front panel of Elimination of 5% program is shown in Fig. 2.

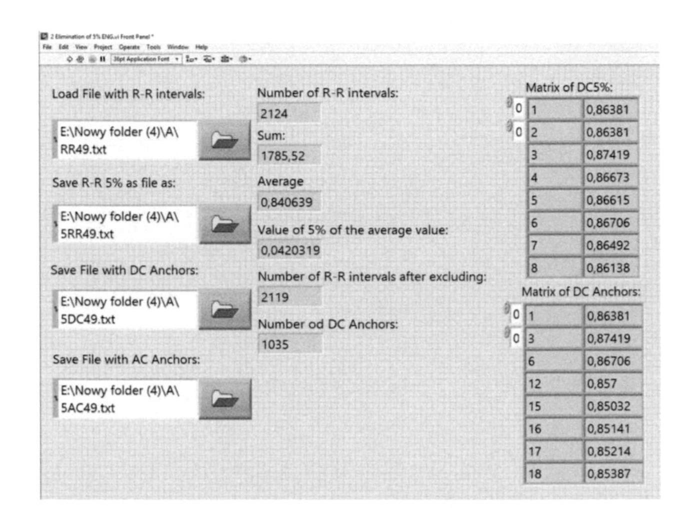

Fig. 2. View of the Front Panel of the Elimination of 5% program in LabVIEW environment.

The AE program after elimination of artifacts designates a new anchors DC and anchors AC and saves them into the files. Program writes data with matrix of R-R intervals as well.

3.3 Assessment of Deceleration Capacity and Acceleration Capacity

The third author's program: Calculation of DC program (CoC) to designate of deceleration capacity and acceleration capacity parameters performs calculations according to the formula [2]:

$$DC(AC) = [X(0) + X(1) - X(-1) - X(-2)]/4 \tag{1}$$

where $X(0)$ = average value of the Anchors DC of the 60th column, $X(1)$ = 61st column, $X(-1)$ = 59th column, $X(-2)$ = 58th column.

Implemented function in the program makes a matrix with saved anchors DC and as well as anchors AC. Front panel of Calculation of DC program is shown in Fig. 3.

The program loads the R-R interval after the elimination and the file from the DC anchors. The calculated value of the DC parameter is presented in the program and saved to the file.

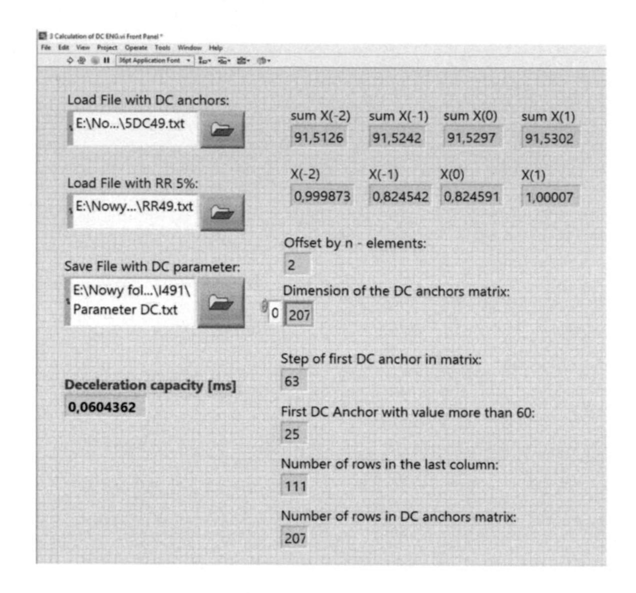

Fig. 3. View of the Front Panel of the Calculation of DC program in LabVIEW environment.

3.4 Statistical Calculations

In order to characterize a series of measurement data properties used the following statistical methods: mean, variance, standard deviation and Student's t-distribution.

Arithmetic Mean of the Measurements. Mathematically, the mean, or average, of N separate values of a sequence x, denoted $Xave$, is defined as [3]

$$Xave = \frac{1}{N}\sum_{n=1}^{N} x(n) = \frac{x(1) + x(2) + x(3) + ... + x(N)}{N} \tag{2}$$

Variance of the Measurements. To designate a variance of a sequence, equation, defined as [3]

$$\begin{aligned}\sigma^2 &= \frac{1}{N}\sum_{n=1}^{N} [x(n) - Xave]^2 \\ &= \frac{[x(1) - Xave]^2 + [x(2) - Xave]^2 + [x(3) - Xave]^2 + ... + [x(N) - Xave]^2}{N}\end{aligned} \tag{3}$$

Standard Deviation of the Measurements. Another measure of a signal sequence is the square root of the variance known as the standard deviation [3]

$$\sigma = \sqrt{\sigma^2} = \sqrt{\frac{1}{N}\sum_{n=1}^{N}[x(n) - Xave]^2} \tag{4}$$

T-test for Independent Groups. To designate a student's t-distribution of DC and AC parameter used equation [4]:

$$t = \frac{\overline{X} - \mu}{S}\sqrt{\nu} \tag{5}$$

where S = a standard deviation of the measurement.

$$S = \sqrt{\frac{1}{n-1}\sum_{i=1}^{n}(Xi - \overline{X})^2} \tag{6}$$

Where n - degrees of freedom.
Student's probability density function given by

$$f(t) = \frac{\Gamma\left(\frac{n}{2}\right)}{\Gamma\left(\frac{\nu}{2}\right)\sqrt{\nu\pi}}\left(1 + \frac{t^2}{\nu}\right)^{-\frac{n}{2}} \tag{7}$$

The above methods have been used for statistical analysis of DC and AC parameters in patients age range 18–58 and 59–60 years.

4 Results

4.1 Collation of AC and DC Parameters of the Subjects

All the values of the measurement of the DC and AC were shown in Table 1. Figure 4 presents a correlation between values of the AC and DC parameters of the subjects.

The table above shows the results of the estimated DC and AC values before (Value of DC and Value of AC) and after elimination of 5% of the interval (Value of DC 5% and Value of AC 5%). In the first and second column is being presented the numbering assigned to patients along with the file name in the PhysioBank database [5]. Next two columns characterize the gender and the age of the patient. V and VII columns describe the DC and AC values established by the CoC program for every patient with exclusion of R-R interval, whose value is about 5% higher of the average. VI and VIII columns describes the values established by DC and AC parameters for every patient without exclusion of any R-R interval. The method of calculation of DC and AC parameters is shown in the Sect. 3.3.

Both AC and DC do not exceed 0.9. The presence of local extremes for the DC factor is more pronounced, while the AC parameter has a lower dispersion of its value.

Table 1. Results of the calculating of deceleration capacity (DC) and acceleration capacity (AC) by cohort.

Patient	File name	Sex	Age	Value of DC 5%	Value of DC	Value of AC 5%	Value of AC
Patient 9	I19.dat	F	18	0,370	0,486	0,389	0,598
Patient 19	I42.dat	M	19	0,027	0,232	0,138	0,056
Patient 16	I35.dat	F	38	0,026	0,398	0,166	0,756
Patient 6	I12.dat	F	39	0,305	0,339	0,130	0,121
Patient 15	I33.dat	M	40	0,281	0,735	0,437	0,766
Patient 14	I29.dat	F	41	0,269	0,019	0,240	0,330
Patient 25	I58.dat	F	45	0,175	0,147	0,165	0,069
Patient 26	I60.dat	F	49	0,849	0,182	0,204	0,148
Patient 11	I23.dat	M	52	0,544	0,259	0,151	0,275
Patient 20	I44.dat	F	53	0,296	0,071	0,066	0,018
Patient 23	I51.dat	M	56	0,018	0,238	0,180	0,119
Patient 7	I15.dat	M	57	0,340	0,366	0,183	0,183
Patient 28	I67.dat	F	58	0,169	0,273	0,183	0,206
Patient 10	I20.dat	F	59	0,113	1,126	0,116	0,716
Patient 13	I27.dat	M	60	0,222	0,146	0,215	0,001
Patient 17	I38.dat	F	60	0,073	0,098	0,230	0,038
Patient 294	s0559_re.dat	F	61	0,219	0,335	0,073	0,048
Patient 8	I17.dat	M	64	0,216	0,212	0,090	0,038
Patient 12	I25.dat	F	66	0,234	0,890	0,035	0,025
Patient 18	I40.dat	M	66	0,132	0,093	0,149	0,118
Patient 32	I75.dat	M	66	0,249	0,061	0,145	0,105
Patient 5	I11.dat	M	68	0,053	2,678	0,177	0,327
Patient 21	I47.dat	F	68	0,232	0,427	0,137	0,546
Patient 22	I49.dat	M	70	0,298	0,062	0,512	0,189
Patient 31	I73.dat	M	73	0,229	0,059	0,150	0,211
Patient 24	I55.dat	M	74	0,258	0,145	0,219	0,178

4.2 Calculation Results

Table 2 shows the results of calculation of three statistical methods: average value, variance and standard deviation for DC and AC parameters in selected age groups.

The calculation was made for the age group of 18–58, 59–74 and the whole population taking part of the research in age of 18–74.

In order to analyze the results of the study, two age groups were compared in terms of DC and AC values. Age groups included women and men between 18

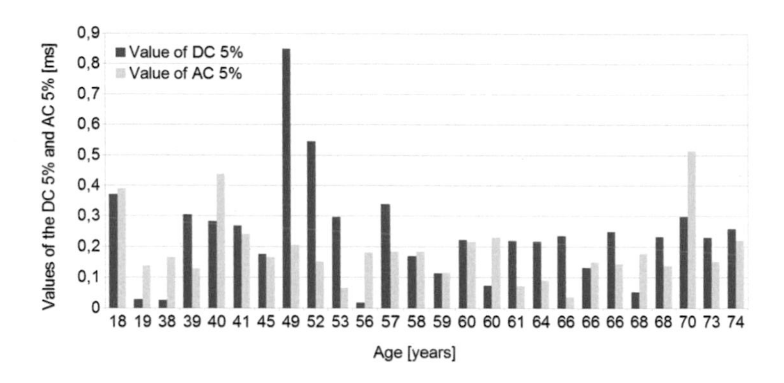

Fig. 4. Collation of AC and DC parameters of subjects.

Table 2. Results of statistical calculations.

Statistical methods	DC [ms]	AC [ms]
Arithmetic mean for subjects in group 18–74 years:	0,238	0,188
Arithmetic mean for subjects in group 18–58 years:	0,282	0,203
Arithmetic mean for subjects in group 59–74 years:	0,194	0,173
Variance for subjects in group 18–74 years:	0,030	0,012
Variance for subjects in group 18–58 years:	0,052	0,010
Variance for subjects in group 59–74 years:	0,006	0,014
Standard deviation for subjects in group 18–74 years:	0,173	0,109
Standard deviation for subjects in group 18–58 years:	0,228	0,102
Standard deviation for subjects in group 59–74 years:	0,076	0,117

and 54 and between 59 and 74 years. Both groups consisted of 13 people. Two mean DC and AC values were obtained.

At this point two hypotheses should be put forth:

- a null hypothesis, which assumes that the mean DC and AC values for the two age groups are equal:

$$p > a$$

- an alternative hypothesis assuming that the mean DC and AC values for the two age groups are significantly different:

$$p \leqslant a$$

Where: p value - calculated correction for a given test
a - significance - significance level assumed at 0,05.

By performing a Student t-test for independent samples, it can be verified whether the mean values of DC and AC for two age groups are statistically significant. The conditions for applying the Student t test are as follows:

- data distribution in groups is similar to the normal distribution
- the variance in groups is equal
- the groups are equal in numbers.

The Kolmogorov-Smirnov test was used to verify whether the results of the estimated DC and AC parameters had normal distribution. A null hypothesis was put forth, assuming that the distribution of the variable is close to normal.

Using PQStat software it was calculated that p-value for Kolmogorov-Smirnov test has a value above the significance level of 0,05. Consequently, the distribution of the variable in the sample is a normal distribution. As a result of the aforementioned assumptions for both age groups for the AC parameter, the probability p was compared with the significance level a (0,05). The value of p was 0,500082, hence the null hypothesis, assuming that the mean value of the AC parameter for the two age groups is equal. In the case of the DC parameter, the variances for the two age groups are not equal, and the condition of the Student t test use is not met. In this case it is possible to use the Mann-Whitney U test (Table 4). This test does not need to meet the conditions of homogeneity of variance and the normality of distribution assumptions. The following hypotheses are presented for Mann-Whitney U test:

- the null hypothesis assuming that there is no difference in the medians (distributions) of the DC parameter between two age groups:

$$p > a$$

- alternative hypothesis assuming that there are differences in the medians (distributions) of the DC parameter between two age groups:

$$p \leqslant a$$

- The value of p is 0,222576 and is compared to the significance level, thus confirming the null hypothesis (Table 4). The results of the normal distribution test are presented in Table 3.

Table 3. Results of normality tests.

Analysed variables	DC	DC	AC	AC
Significance level	0,05	0,05	0,05	0,05
Data Filter	Group 18-58	Group 59-74	Group 18-58	Group 59-74
Group size	13	13	13	13
Group mean	0,282231	0,194462	0,202462	0,172923
Group standard deviation	0,228512	0,075842	0,102424	0,117059
Kolmogorov-Smirnov test				
D statistic	0,196609	0,3041	0,267657	0,235996
Degrees of freedom	13	13	13	13
p-value	0,627792	0,145597	0,259311	0,401383

Table 4. Mann-Whitney U test for DC parameter.

Analysed variables	DC;Age group
Significance level	0,05
Continuity correction	Yes
Grouping variable	Age group
Group name	Group 18-58
Group size	13
Sum of the ranks for group	200
Mean of the ranks for the group	15,384615
Group median	0,281
Group name	Group 59-74
Group size	13
Sum of the ranks for group	151
Mean of the ranks for the group	11,615385
Group median	0,222
U statistic	60
U statistic	109
p-value (exact)	0,222576
Z statistic	1,230769
p-value (asymptotic)	0,218409

Table 4 presents the results of the Mann-Whitney U test for DC parameter. Normal distribution plots are shown in the Figs. 5, 6, 7, 8.

Table 5 presents the results of the T-student test for the two age subgroups: 18–58 and 59–74 years for the AC parameter.

Plots (Figs. 9 and 10) show the medians for the DC and AC parameters.

Due to the nature of data distribution in the test, which is similar to normal, it was possible to perform Student t-test for the AC parameter and the Mann-Whitney test for the DC paramcter.

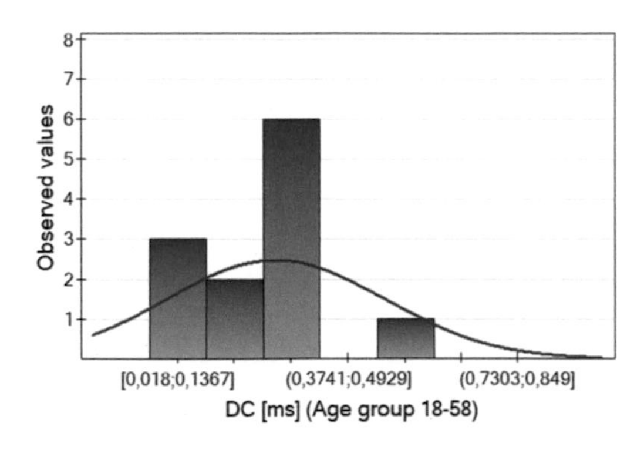

Fig. 5. Plot of the normal distribution for DC parameter for age group 18–58.

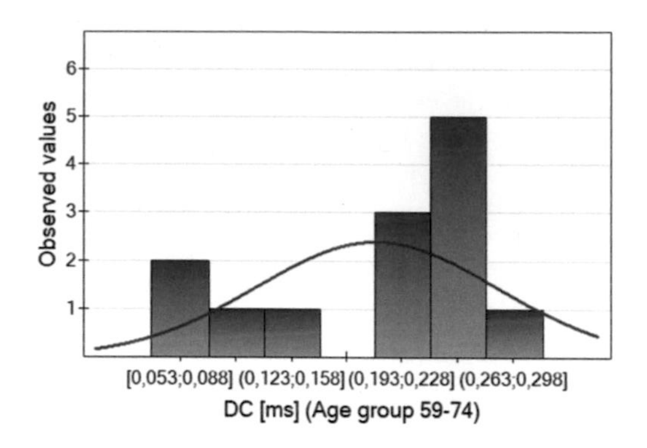

Fig. 6. Plot of the normal distribution for DC parameter for age group 59–74.

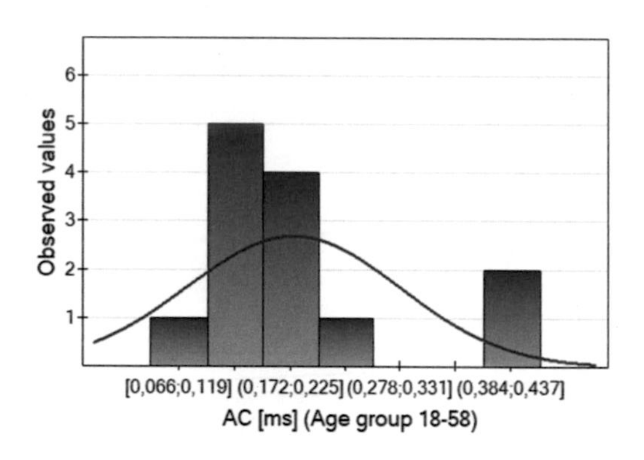

Fig. 7. Plot of the normal distribution for AC parameter for age group 18–58.

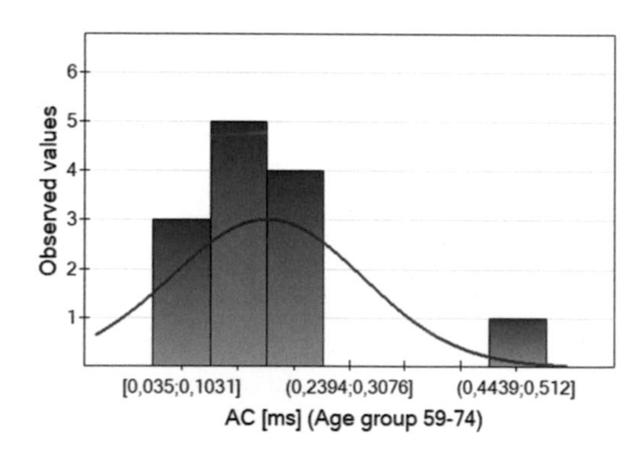

Fig. 8. Plot of the normal distribution for AC parameter for age group 59–74.

Table 5. T-test for AC parameter.

Analysed variables	AC 5% [ms];Age group
Significance level	0,05
Correction for different variances	No
Grouping variable	Age group
Group name	18-58
Group size	13
Group mean	0,202462
Group standard deviation	0,102424
-95% CI for the group mean	0,140568
+95% CI for the group mean	0,264356
Group name	59-74
Group size	13
Group mean	0,172923
Group standard deviation	0,117059
-95% CI for the group mean	0,102185
+95% CI for the group mean	0,243661
Difference of the means	0,029538
-95% CI for the difference	-0,059497
+95% CI for the difference	0,118574
Standard error of the difference	0,04314
Pooled standard deviation	0,109985
t-statistic	0,684718
Degrees of freedom	24
Two sided p-value	0,500082
Fisher-Snedecor test	
Variance ratio F	0,765584
p-value	0,650927

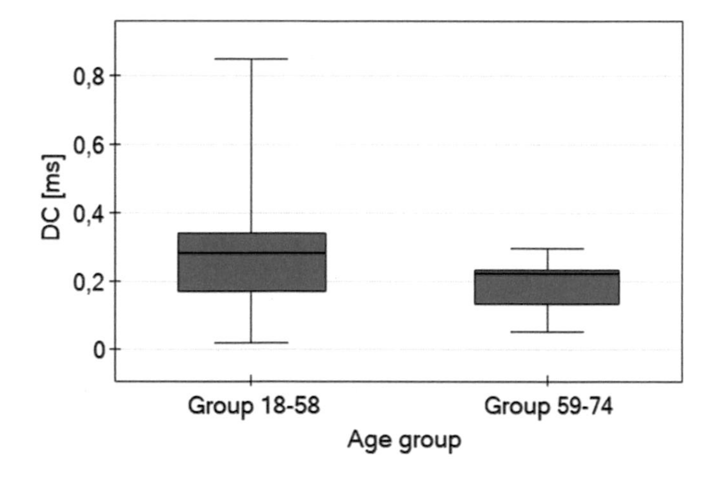

Fig. 9. Plot of the median group for DC parameter.

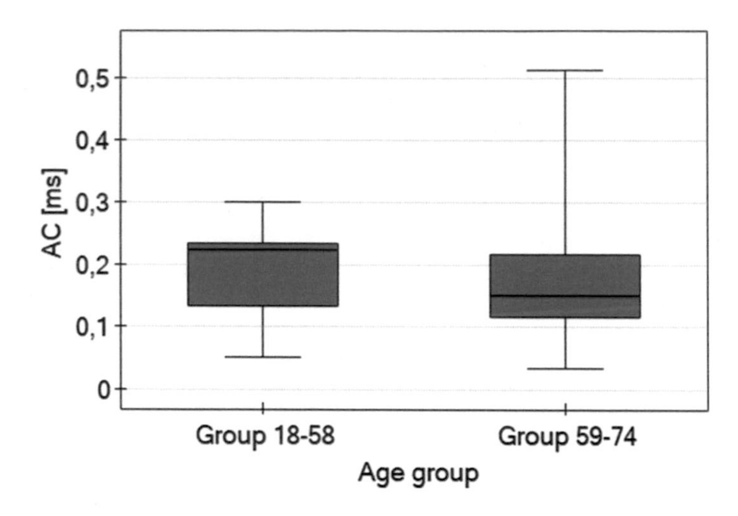

Fig. 10. Plot of the median group for AC parameter.

The statistical analysis indicated no significant differences between the mean values of DC and AC in the two age groups. It can be concluded that the patient's age has no influence on the DC and AC parameters and their value mainly depends on the cardiovascular status of the patient.

As a result of the study performed on patients in London and Oulu, the DC parameter was assigned to a three groups of mortality risk [2]. The proposed boundaries for DC parameter: DC parameters for high-risk group (deceleration capacity $\leqslant$ 2,5 [ms]), intermediate-risk (2,6 to $\leqslant$ 4,5 [ms]), and low risk (> 4,5 [ms]) [2]. In the test the calculated arithmetic mean for subjects in group 18–58 years gives value 0,28 [ms] and for subjects in group 59–74 years 0,19 [ms]. The DC parameters specified in this article are considered to be expected in view of the fact that DC values are less than 2,5 [ms] and the study was performed on a group of patients after myocardial infarction with the high-risk of mortality. The data analysis has included statistical analysis of DC and also AC parameter. In the examination case study used the T-Student test (AC) and Mann-Whitney U test (DC) for independent variables. As the results of the study it was proved the null hypothesis, that assumes that the patient's age after myocardial infarction does not affects the value of the DC and AC parameters.

5 Conclusion

This article presents three author programs designed in the LabVIEW environment for electrocardiographic analysis. For research purposes, we used ECG signals from the free PhysioNet database. Using the LabVIEW Biomedical Toolkit, one can determine QRS complexes from a digital heart beat record, and then estimate the DC and AC parameters. This article focuses on the analysis of statistical data, calculating such measures as mean, variance, or standard deviation. Student t-test was also used for data analysis for DC and AC parameters in the age range of 18–58 and 59–74 years. According to the results of Kolmogorov-Smirnov test, it was proved that the data have normal distribution. Student t-test showed that the null hypothesis assuming that the mean value of the AC parameter for the two age groups does not differ significantly. For the DC parameter, the Mann-Whitney test confirms the hypothesis that there is no difference in the median DC values between the two age groups. Estimated results of the DC and AC parameters were performed for a small population of only 26 persons. In addition, it was observed that the digital data contained in the PhysioNet database had numerous errors in the form of artifacts, so that the LabVIEW Biomedical Toolkit was having trouble identifying all the R segments. The subject is developmental and the study of DC and AC parameters will be continued using the appropriate medical equipment including ECG holter with advanced RR interval determining software.

References

1. Sacha, J., Sobon, J., Sacha, K., Muller, A., Schmidt, G.: Short-term deceleration capacity reveals higher reproducibility than spectral heart rate variability indices during self-monitoring at home. Int. J. Cardiol. **152**, 271–272 (2011)
2. Bauer, A., et al.: Deceleration capacity of heart rate as a predictor of mortality after myocardial infarction: cohort study. Lancet **367**, 1674–1681 (2006)
3. Lyons, R.G.: Understanding Digital Signal Processing. Angewandte Chemie International Edition 40, p. 665. Prentice Hall, Upper Saddle River (2004)
4. The Student t Distribution: Basic Theory. http://www.math.uah.edu/stat/special/Student.html
5. PhysioNet: A Resource for Research and Education. https://physionet.org/cgi-bin/atm/ATM

Convolutional Neural Networks Implementations for Computer Vision

Paweł Michalski[1], Bogdan Ruszczak[2], and Michał Tomaszewski[1](✉)

[1] Faculty of Electrical Engineering, Automatic Control and Informatics, Institute of Computer Science, Opole University of Technology, Prószkowska 76, 45-758 Opole, Poland
{p.michalski,m.tomaszewski}@po.opole.pl

[2] Faculty of Economy and Management, Luboszycka 7, 45-036 Opole, Poland
b.ruszczak@po.opole.pl

Abstract. The paper covers the current state of the art regarding the use of machine learning mechanisms, and in particular the deep convolutional neural networks used in the field of computer vision. In the article there has been presented the current definition of deep learning and specific dependencies between related fields such as machine learning and artificial intelligence. The practical part of the work consists of three components: the features of the structure of the convolutional neural network, the distinction of its key elements, the description of their actions, the compilation of information about available learning sets used in network testing and verification processes, and the review of the implementation of convolutional neural networks, which had a significant impact on development of discipline. To illustrate the great potential of the presented tools for solving computer vision tasks, the study highlites examples of their applications. The possibility of using convolutional neural networks for identification of technical objects in digital images is indicated.

Keywords: Deep learning · Convolutional neural networks
Computer vision · Artificial intelligence · Machine learning

1 Introduction

Processing the large data sets (big data) is currently a rapidly growing discipline due to increasing demand in a number of areas, such as medical data processing [1], supporting the movement of autonomous cars [2], or the processing of the digital images [3]. Machines, as a tool to replace people's work, are now equipped with sets of sensors that respond to human senses. In the case of people, the sense of sight is one of the most important senses, and provides a great deal of information about the surroundings. Despite the fact that for a long time there

Paweł Michalski, PhD. Eng., Assistant Professor; Bogdan Ruszczak, PhD. Eng., Assistant Professor; Michał Tomaszewski, PhD. Eng., Associate Professor.

W. P. Hunek and S. Paszkiel (Eds.): BCI 2018, AISC 720, pp. 98–110, 2018.
https://doi.org/10.1007/978-3-319-75025-5_10

has been a wide range of vision sensors available to record images in various electromagnetic fields, it is still troublesome to process the vision information in the shortest time possible, especially since many of these tasks require real-time analysis. The autonomous machine that performs the tasks required should very often have the ability to determine its position relative to the surroundings, such as furniture, traffic congestion in rooms or other road users [4,5]. Machines, like people, must make a specific decision based on acquired images. Another issue is to identify elements of the environment, which is widely analyzed by researchers around the world [6].

Another task, pattern recognition, which for most people is not difficult, in the case of computers turns out to be a much more complicated process. This is most often caused by the necessity to map 3D objects with a single image or a set of two-dimensional images. Additionally, the recording of images can be done in different lighting conditions, which also affects the degree of difficulty of the process. Another difficulty may be partial occlusion of the subject or image projection from another perspective as compared to the master image. Despite the differences in lighting and in spite of the various points of view, the task of classifying a facility by man in most cases is still successful. In the case of automated processes performed by the machines it is required to use one of the common patterns matching algorithms, such as Scale Invariant Feature Transform (SIFT), Speed Up Robust Feature (SURF) or Robust Independent Elementary Features (BRIEF). Some variants of these algorithms, for example Oriented FAST or Rotated BRIEF (ORB) can accurately detect a pattern with deformations such as the change of scaling or rotation [7]. The key to a good object recognition is the set of features which describe it. Such features are then sought for in test images, and on this basis a set of features, describing the object being tested, is created.

Classic Computer Vision (CV) algorithms unfortunately have their limitations and usually get the correct results with small deformations of objects, and under similar lighting conditions, which were associated with the building of a model set of characteristics describing an object. Widespread use of Convolutional Neural Networks (CNN) as a tool to aid the machines to make decisions based on a set of images of the surrounding world was possible through two processes. The first was the appearance of CPUs with high computing abilities, especially high-performance graphics cards processors which do very well in the network learning process. The second was the development of Machine Learning (ML) techniques, in particular Deep Learning (DL) algorithms.

Both ML and DL are concepts in the field of research of Artificial Intelligence (AI) or Computational Intelligence (CI). The machine learning process can be divided into two main groups: unsupervised learning and supervised learning, where the machine learns based on test data. The test data usually comes in the form of input/output pairs, written to give you basic information to make learning the future decisions of the system. Such a learning process involves the participation of man in the learning process. During supervised learning process for each input information must first be assigned a correct response at the output of the circuit. It's not always possible to use this automatic learning method, then the use of unattended teaching remains an alternative. DL is implemented by large-scale neural networks, which the best example are the deepest convolutional neural networks described in this paper.

The rapid development of AI techniques is measured with some troubles, resulting from the activity of the proposed algorithms. In the discussion initiated by the authors in [8], the main limitations of AI were identified. During the processing large collections of learning, it's an difficult task to automatically identify costly and often repetitive activities in the process of searching for a solution. With limited information, the algorithm has no basis for automatically eliminating certain sets of solutions, even if it might seem pointless to humans. One of the limitations is also the insufficient skill of AI algorithms to adapt to changing conditions. Their work runs smoothly if they are powered with data for which they have been prepared, or more broadly, to solve even complex problems but with the right structure.

2 Deep Learning

DL is based on in-depth AI research and is part of ML, and was created during the explosion of popularity of AI-related IT tools. The illustration below (Fig. 1) shows technologies sometimes called "narrow AI" – they perform precisely defined tasks with similar efficiency as humans, and sometimes even better. Examples of such tools can be automatic classification of photos in Pinterest or face recognition in photos posted on Facebook.

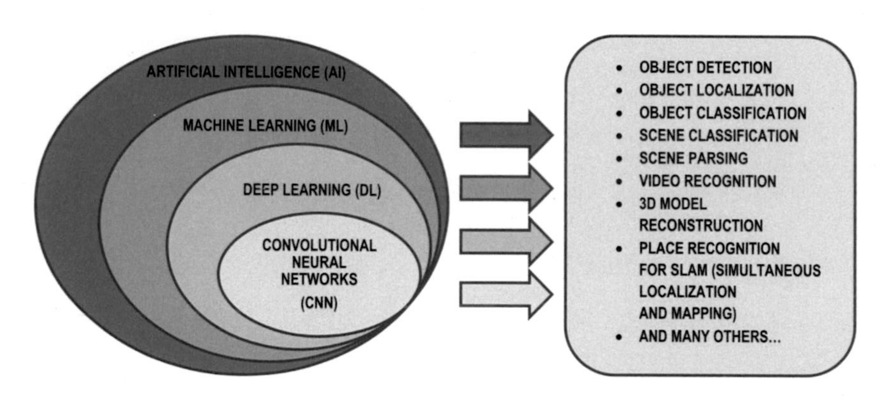

Fig. 1. Relationships between artificial intelligence, machine learning, deep learning and convolutional neural networks and examples of applications in computer vision.

In the simplest sense, ML is a process that comprises of parsing data, automatic learning based on them, and subsequent results determination or prediction estimation. ML is rather training to perform specific tasks using a large set of data than manually algorithms coding to solve problems.

ML is an attempt to tackle the challenges of AI at the beginning of its development, which have been tried before with, for instance: decision tree learning, clustering, reinforcing learning, Bayesian networks, inductive logic programming, etc. Those methods allowed to perform a narrow range AI only. Nowadays researchers try to solve these issues using ML. One of the first major projects in ML was Google Brain, which created the TensorFlow library, which is one of the popular tools for creating and teaching neural networks, published under the open Apache license. Another interesting use of deep learning in practice is the voice recognition implemented in applications like Google Now and Apple Inc.'s Siri.

DL has enabled the practical solution of many tasks where ML was not enough. Issues such as the navigation of autonomous cars, recommendations for consumers in sales systems or support for preventive health care are currently being implemented by the DL. One of the areas where using such algorithms is nowadays broadly implemented is CV.

3 Convolutional Neural Networks as a Technique for Deep Learning

According to [9], Deep Learning is a section of Machine Learning, centered around algorithms modeling high-level abstraction in data sets, using multiple layers of nonlinear transformations. The most well-known and used group of deep learning algorithms are Convolutional Neural Networks because of their wide application possibilities in recognizing different patterns, particularly in detecting objects in digital images.

The idea of CNN has been around for a long time, but there were many problems that inhibited its development. For example, with the enlarging number of network's layers, the number of model parameters increases, which, in conjunction with the simultaneous rise in the size of the input training base, significantly expands the demand for computing power. The vanishing gradient problem, associated with the use of the reverse propagation algorithm, also required a solution.

In recent years there has been a rapid development of CNN, and the impact of many difficulties has been significantly reduced as a result of research into new concepts for the broader neural network, including [9–16]. At the same time, there has been an increase in the ability to create very large datasets [17] based on a big data revolution.

4 Components of the Convolutional Neural Networks

What are Convolutional Neural Networks? In the construction of these networks, there are four main types of layers that perform the basic tasks of such networks: convolution, pooling, normalization and connection. As a result of the Convolution Layer, the input image is processed by a variety of convolutional filters to extract the characteristics contained in those parts. Pooling Layer is being use for reduce the size of the information being analyzed, thereby decreasing the sensitivity of the network to the distortion of the analyzed scene. The basic methods used in this layer are max pooling, when the largest value is selected in the parsed window and averaging, when its value is averaged. The ReLED layer (Rectified Linear Units Layer) by data normalisation increases the network's ability to solve nonlinear problems. CNN consist of multiple layers on successive levels, but the last link in such a network is the submission of results to the final layer – Fully Connected Layer. This layer results in the final rating, allowing the various training categories assignment.

The distinguishing feature of CNN over classical neural networks is that the number of layers is much higher. The Fig. 2 shows the structure of CNN. The depth of neural network architecture is defined as the length of the longest path between the input and output neurons. There is no precise threshold of the layers number, allowing one to call the network "deep", but it has been assumed that it refers to the network with more than two hidden layers.

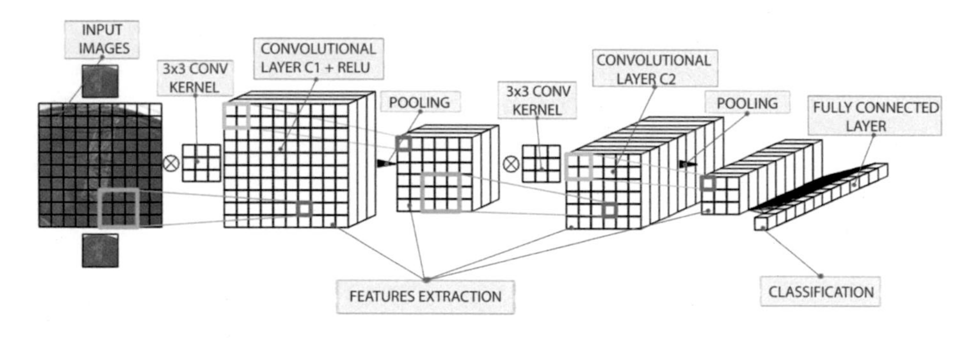

Fig. 2. The general structure of the convolutional neural network with distinction of the most essential components.

The general structure of the CNN described above can be modified. An alternative to the ReLU layer may be to use the Softmax or Parametric Rectified Linear Units (PReLU) function [16], which could improve the network structure. Another interesting approach to CNN modification using the parallel processing of several convolutional layers (Inception Module) is presented in [18].

The process of designing a CNN requires matching its structure - determining the number of layers and their respective layout, as well as the parameters of their work. The preparation of each layer consists in defining the so-called hyperparameters, such as depth defining the number of neurons (different type and size features or filters on each of the Convolutional Layers), stride deciding how dense sampling of layers will be performed, zero-padding specifying how to supplement incomplete information (eg. on the edges of the processed images), and the number of neurons in the Fully Connected Layer.

The CNN learning involves processing input datasets (e.g. image collection) with assigned categories. This is done in unattended mode. The mechanism used to train the network that distinguishes the convolutional neural networks is the error backpropagation. This method allows to improve the grading scales of the individual layers based on the observation of the adjustment evaluation error function. The purpose of the operation is to modify the weight of the classifier so that the observed adjustment error is lower. The phenomenon of the overfitting CNN is also worth mentioning. Especially Fully Connected Layer, which stores most of the information for the network, is susceptible to this treat. A special technique, called dropout, allows to stop sub-elements learning before overtraining the entire network.

5 ImageNet as a Source of Data for Convolutional Neural Network

Since the widespread use of digital documents and the availability of global data interchange (internet development), digital imaging specialists have worked to design more sophisticated algorithms for indexing, downloading, organizing, and commenting on multimedia data. CNNs learning to recognize specific objects in digital images requires a large number of digital images to be stored in a database which must be catalogued according to a specific hierarchy. This led to the idea of creating a large, indexed database of digital images - datasets. The most commonly used database for this purpose is the ImageNet [19] project, which is designed to conduct research to identify a variety of objects. ImageNet is a collection of digital images organized according to the hierarchy used in the WordNet dictionary. In this Princeton University dictionary, every significant term which is possible to be described by many words or phrases is called a "set of synonyms" or "synset". Today there are more than 100,000 synonyms in the WordNet glossary, most of them nouns (80,000+).

Table 1. ImageNet database (April 30, 2010) [19].

Total number of non-empty synsets	21,841
Total number of images	14,197,122
Number of images with bounding box annotations	1,034,908
Number of synsets with SIFT features	1000
Number of images with SIFT features	1,200,000

Based on dependencies defined in WordNet, nearly 15 million URLs of digital images have been added to ImageNet by 2016, which were manually described to name the objects. Additionally, over a million digital images have bordered named objects. There has been created a giant set of digital images used by scientists from around the world. Table 1 shows selected statistics of ImageNet database.

Currently ImageNet is widely used by scientists around the world in developing new methods in the field of CV, particularly as a training material for CNNs. Since 2010, ImageNet has been organizing the annual ImageNet Large Scale Visual Recognition Challenge (ILSVRC), where teams of researchers create competitive software in the field of valid classification and object and scene detection, based on ImageNet. The Contest is a natural successor to the Caltech 101 and PASCAL VOC projects, but on a much larger scale - for comparison, the PASCAL VOC image database in 2012 contained only about 21,738 images grouped in 20 classes.

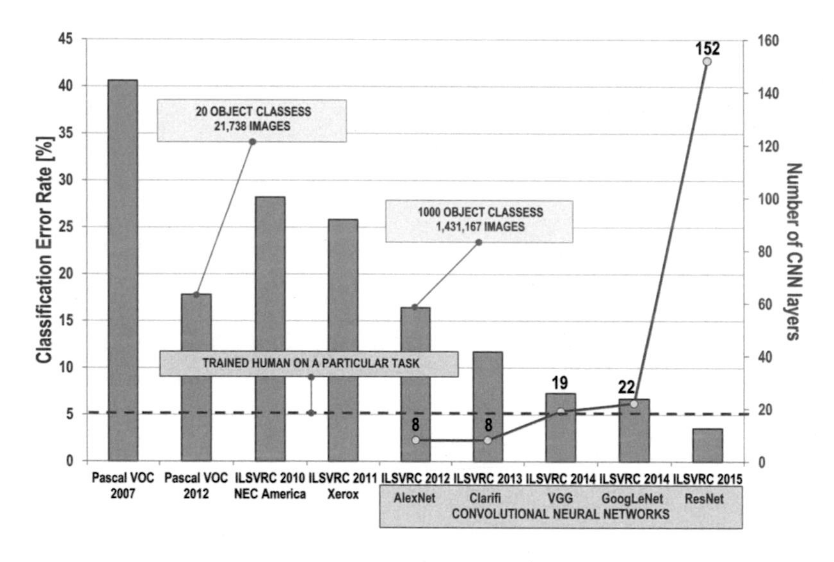

Fig. 3. Achievements in image classification. Based on: [17, 19–23].

With each new contest and the growth of the ImageNet database, there has been a sharp increase in the processing of digital images year after year (Fig. 3). In 2011, the ILSVRC rate of error was 25%. Compared to the winner of the Pascal VOC the average error rate was 40.65% in 2007 [22] and 17.8% in 2012 [23] respectively. In the following years, the development of digital image processing techniques for analysis and the use of CNNs have reduced error rates in subsequent years to a few percent. It was announced in 2015 [17] that the software outperformed human capabilities in narrow image analysis tasks from ImageNet. ResNet, the winner of the ILSVCR in 2015, achieved 4.8% during, and 3.57% after the competition.

Recognizing the positive results of the research that has been made with the ImageNet database, other similarly developed data sets have recently been launched. For example, DigitalGlobe has released a free of charge SpaceNet [24] repository of high resolution satellite imagery to support the development of new remote sensing applications (automatic recognition and extraction of satellite imagery). ImageNet is currently the largest categorized database of digital images, but it contains images in low resolution. Data provided in the SpaceNet repository will primarily help in developing computer vision algorithms for automatic building detection, but the authors believe that they can also be used to identify technical objects (power lines, transport infrastructure, industrial networks). In the near future, SpaceNet's repository will offer far greater possibilities, with DigitalGlobe announcing it will include as many as 60 million satellite scenes.

As building large (massive) datasets is time consuming and costly, a number of work is being done to optimize these activities. Recent interest in Generative Adversarial Networks (GAN) has been raised in [15]. These are methods that could be used to create data sets that are similar to input data. In order to approximate the concept of the GAN, the distinction between generative models and discriminatory models must first be indicated.

The discriminant model in the learning process fits a function that assigns the input data (x) to the desired category (y). Under probabilistic circumstances, the conditional distribution $P(y|x)$ is directly taught. In this way, extended hierarchical models [13] are created that represent the probability distributions of data processed by deep CNN. The generative model tries to learn simultaneously the probability of a set of input and category data simultaneously, e.g. $P(x, y)$. This probability can be converted to $P(y|x)$, respectively, for assigning an object to a set, using the Bayes' theorem, but additionally generative capabilities can also be used to create new sets of data (x, y).

GAN implementation is based on two models of neural networks competing with each other. One model gets unclassified input and generates samples (generator). The second model (discriminator) receives samples from both the generator and the input learning data, and must be able to distinguish between the two data sources. These two networks play a continuous game of fixed sum, where one player's profit is the loss of the other. During the game the generator learns to produce more realistic samples, and the discriminator learns better and better distinguishes the generated data from the real data. The most important benefit of GAN is that we can provide information about the backpropagation gradient from the generator to discriminator network. Both types of models are useful, but generative models have one interesting advantage over discriminatory models - they can understand and explain the basic structure of input data even when they are not assigned to any category. This is particularly important when working on real-world data modeling, where uncategorized data is very extensive, and the acquisition of the described data is expensive and often impractical.

6 Selected Examples of Convolutional Neural Networks

LeNet

CNNs are a type of network that try imitate the human perception of the environment. It is the first stage in the processing of information that is received through the senses. The originator of the concept of processing information in a similar manner by machines was a scholar, Frank Rosenblatt, a psychologist who studied the processes of animal learning. He was also the creator of the first image recognition system which he called the perceptron. Minsky and Seymour Papert some years later showed limitations on perceptrons that temporarily inhibited work in this field [25]. Another important discovery was the development of neocognitron, which was developed as a handwriting recognition system [26]. The first significant CNN was the LeNet-5 network created by Yann LeCun. LeNet-5 had a multilayered structure and allowed the use of backward error propagation algorithm in the network learning process [10]. The network was designed to recognize handwriting and print and was able to correctly classify characters after tampering (inter alia translation, scale, rotation, squeezing, stroke). As input, a 32×32 pixel image was provided. The network structure consisted of 6 layers (3 convolutional layers, 2 subsampling layers, 1 fully connected layers). The network was learned using the MINIST dataset.

AlexNet

The AlexNet is the implementation that achieved the highest score in the ILSVRC test in 2012, obtaining a Top5 test error of 16.4% (Top5 error is an indicator of whether the search object was in the one out of 5 best-fit categories). The author of the network is Alex Krizhevsky, his network is based on 5 convolutional layers, max pooling, dropout, and 3 fully connected layers [14]. The network structure uses also the Rectified Linear Unit (ReLU) activation function. Input to the first convolutional layer is given in the form of a matrix of $224 \times 224 \times 3$. Filters used in the first layer have a dimension of 11×11. The network has been pre-trained using the ImageNet database and allows classifying objects for 1000 possible classes. Network learning was done using the stochastic gradient descent method. The network has 650 K neurons and using 630 M connections generates 60 million parameters, so its authors have devised a dropout mechanism to remove some neurons and their connections during the learning process. Implementation of ReLU activation with dropout during the network learning process let one to skip the pre-training phase when very large amount of labeled data is available [27]. The AlexNet learning lasted 5 days with 2 GTX 580 graphics cards.

ZF NET

Another network to focus on ZF Net was developed by two researchers, Matthew Zeiler and Rob Fergus in 2013 and it get a Top5 error rating of 11.2% in ILSVRC.

The network has been developed on the basis of the AlexNet with several modifications. The main advantage is the reduction of the filter window to 7×7, resulting in much more information concerning the original image and the use of the activation functions: cross-entropy and the loss for the error function. The network was taught using the GTX 580 for twelve days. The most interesting component of the network was the deconvolutional layer that could reverse the convolution process, resulting i.e. in the ability to visualize activation of neurons in intermediate layers. The deconvolutional network, using grouping and regularization, in a reversed way, allow to obtain the corresponding pixels to the input image [28]. The visualization mechanism, developed in ZF Net, could be used to preview how each layer works, and allows to introduce network architecture improvements.

VGG

The VGG Net was created in 2014 by the Visual Geometry Group team, which, unlike the previously mentioned networks, distinguishes the increased number of layers used. The network structure was originally developed in several variants ranging from 11 up to 19 layers. The aim of this approach, developed by Karen Simonyan and Andrew Zisserman, was the idea of to investigate how the depth of the network affects the results [29]. The main difference with respect to earlier networks is the use of 3×3 filters, in comparison AlexNet uses 11×11 and ZF Net 7×7. Consecutive convolution of two 3×3 filter layers yields an effective 5×5 receptive field and a combination of three 3×3 convolution layers results in a effective 7×7 receptive field. This approach allows to maintain benefits of a large filter with the use of several smaller ones and to reduce the number of parameters to learn. The network was learned using the mini-batch gradient descent method.

GoogLeNet

GoogLeNet is the network that won the ILSVRC competition in 2014 and achieved 6.7% error rate in the Top5 category. The network developed by Google has a different structure than the previously discussed. It consists of 22 layers so the trend of deepening the network structure (shown in Fig. 3) is preserved. The increasing number of layers causes the raise in the number of parameters and could lead to overfitting and the computational cost growth. The GoogLeNet network generates up to 12 times less parameters than AlexNet, and while being significantly more accurate. Such optimization has been achieved by introducing a mechanism called the Inception Module, what is an important enhancement over the sequential approach of the remaining networks. In the previous implementations of the CNNs it was necessary to choose whether to perform a pooling or a convolution operation, and to determine the filter size. The Inception Module allows the parallel use of different size filters [18]. For example, it distinguishes general (5×5) and local (1×1) filters at the same time. Then the concatenation of all results into a single vector, used as the next layer input, has to be performed. Overall, the GoogLeNet architecture comprises of a total of over 100 layers, 9 of which were the inception modules. The network structure is currently being developed under the codename Inception, with the most up-to-date version called Inception V4 [30].

ResNet

ResNet was developed by a Microsoft Research Asia team in 2015. It comprises of 152 layers and in the ILSVRC 2015 competition it got an error rate of 3.57%. This result is spectacular, it was the first time when AI achieve better results than human who in the image classification task, depending on their knowledge and experience, oscillate between 5 and 10%. The researchers while defining ResNet found out that the shallow structures are less error-prone during the learning process due to use of residual transition between layers [31]. The concept of residual transition assumes that after passing through several layers, the result is summed up with the input of the CNN. This results in faster learning process of the network. Another change from previous approaches is the abandonment of the fully connected layer to the global average pooling in order to simplify the relation between feature maps and categories, what makes the network results more meaningful and interpretable.

7 Summary and Conclusions

The progress in the classification of objects in digital images by CNN is significant, exceeding even human capabilities. However, it is important to remember that deep learning algorithms can classify only images belonging to strictly defined set of categories. People can recognize a much larger number of classes, create a new ones, while simultaneously being able to analyze the context of the scene [17]. Despite the significant development of CNN there are still some limitations to their common application. In order to achieve higher universality of current solutions the learning mechanisms has to be developed. Many ongoing CNN work is currently centered around the implementation of this tool for solving further Computer Vision tasks. Current editions of ILSVRC competition are focused on object detection, scene classification and scene parsing.

The authors now recognize the wide potential for the development of specialized applications of automated inspection methods, mainly for objects that have not yet become a part of training datasets. The CNNs seems to be suitable to use with technical facilities and equipment, such as power infrastructure. The monitoring systems in this sector already collect the vast amount of data and this brings the need of automated tools introduction to process them. Preparing datasets for new domains and advancing analysis mechanisms on its basis may be necessary. Deep learning mechanisms require very large sets of data, but their preparation requires a considerable effort, it is time and cost consuming, since the assignment of categories to specific objects is still performed manually.

Currently, the work is being done on optimizing these activities, such as the minimizing required learning sets size, while the maintaining the same effectiveness of the trained networks. In order to achieve this the current learning algorithms improvement is essential, for instance using GAN.

The present challenge for the CNN development is the proposal of their new structures (of higher number of layers or enhanced architectures). Network hyper-parameters such as: depth, the number of filters in each convolution, and their size, are correlated, and their fitting requires later verification. A lot of work is currently underway to optimize this process (inter alia [9,16,18,29,32]) and the authors anticipate the appearance new deep learning solutions.

References

1. Batra, S., Sachdeva, S.: Suitability of data models for electronic health records database. In: Srinivasa, S., Mehta, S. (eds.) BDA 2014. LNCS, vol. 8883, pp. 14–32. Springer, Cham (2014). https://doi.org/10.1007/978-3-319-13820-6_2
2. Bagloee, S.A., Tavana, M., Asadi, M., et al.: Autonomous vehicles: challenges, opportunities, and future implications for transportation policies. J. Mod. Transport. **24**(4), 284–303 (2016). https://doi.org/10.1007/s40534-016-0117-3
3. Pal, S.K., Meher, S.K., Skowron, A.: Data science, big data and granular mining. Pattern Recogn. Lett. **67**(2), 109–112 (2015). https://doi.org/10.1016/j.patrec.2015.08.001
4. Häne, C., Sattler, T., Pollefeys, M.: Obstacle detection for self-driving cars using only monocular cameras and wheel odometry. In: IEEE/RSJ International Conference on Intelligent Robots and Systems, IROS. Hamburg (2015). https://doi.org/10.1109/IROS.2015.7354095
5. Salman, Y.D., Ku-Mahamud, K.R., Kamioka, E.: Distance measurement for self-driving cars using stereo camera. In: Proceedings of the 6th International Conference on Computing and Informatics, ICOCI 2017, Kuala Lumpur (2017)
6. Hohm, A., Lotz, F., Fochler, O., Lueke, S., Winner, H.: Automated Driving in Real Traffic: from Current Technical Approaches towards Architectural Perspectives. SAE Technical Paper (2014)
7. Karami, E., Prasad, S., Shehata, M.: Image matching using SIFT, SURF, BRIEF and ORB: performance comparison for distorted images. In: Newfoundland Electrical and Computer Engineering Conference, IEEE, Newfoundland and Labrador Section At St. John's, NL (2015). https://doi.org/10.13140/RG.2.1.1558.3762
8. Amodei, D., Olah, C., Steinhardt, J., Christiano,,P., Schulman, J., Man, D.: Concrete Problems in AI Safety (2016). arxiv.org/abs/1606.06565
9. Goodfellow, I., Bengio, Y., Courville, A.: Deep Learning. MIT Press, Cambridge (2016)
10. LeCun, Y., Bottou, L., Bengio, Y., Haffner, P.: Gradient-based learning applied to document recognition. IEEE **86**(11), 2278–2324 (1998)
11. Hochreiter, S., Bengio, Y., Frasconi, P., Schmidhuber, J.: Gradient flow in recurrent nets: the difficulty of learning long-term dependencies. In: Kremer, S.C., Kolen, J.F. (eds.) A Field Guide to Dynamical Recurrent Neural Networks. IEEE Press, Hoboken (2001)
12. Hinton, G.E.: To recognize shapes, first learn to generate images. Prog. Brain Res. **165**, 535–547 (2007)

13. Bengio, Y.: Learning Deep Architectures for AI. Now Publishers, Boston (2009)
14. Krizhevsky, A., Sutskever, I., Hinton, G.E.: ImageNet classification with deep convolutional neural networks. In: Advances in neural information processing systems (2012)
15. Goodfellow, I., Pouget-Abadie, J., Mirza, M., Xu, B., Warde-Farley, D., Ozair, S., Courville, A., Bengio, Y.: Generative Adversarial Networks (2014). arxiv.org/abs/1406.2661
16. He, K., Zhang, X., Ren, S., Sun, J.: Delving deep into rectifiers: surpassing human-level performance on ImageNet classification. In: Proceedings of the IEEE International Conference on Computer Vision (2015). arxiv.org/abs/1502.01852
17. Russakovsky, O., Deng, J., Su, H., et al.: ImageNet large scale visual recognition challenge. Int. J. Comput. Vis. **115**(3), 211–252 (2015). https://doi.org/10.1007/s11263-015-0816-y
18. Szegedy, C., et al.: Going deeper with convolutions. In: Proceedings of the IEEE Conference on Computer Vision and Pattern Recognition (2015)
19. ImageNet Project. http://image-net.org
20. Cao, J., et al.: A parallel Adaboost-Backpropagation neural network for massive image dataset classification, Sci. Rep. 6(38201) (2016). https://doi.org/10.1038/srep38201
21. Oquab, M., Bottou, L., Laptev, I., Sivic, J.: Learning and transferring mid-level image representations using convolutional neural networks. In: IEEE Conference on Computer Vision and Pattern Recognition, Columbus, OH (2014). https://doi.org/10.1109/CVPR.2014.222
22. Marszalek, M., Schmid, C., Harzallah, H., Weijer, J.: Learning object representations for visual object class recognition. In: Visual Recognition Challange workshop, ICCV (2007)
23. Yan, S., Dong, J., Chen, Q., Song, Z., Pan, Y., Xia, W., Huang, Z., Hua, Y., Shen, S.: Generalized hierarchical matching for sub-category aware object classification. In: Visual Recognition Challenge workshop, ECCV (2012)
24. SpaceNet. http://explore.digitalglobe.com/spacenet
25. Papert, S., Minsky, M.: Perceptrons: An Introduction to Computational Geometry. MIT Press, Cambridge (1988)
26. Fukushima, K.: Neocognitron: a self-organizing neural network model for a mechanism of pattern recognition unaffected by shift in position. Biol. Cybern. **36**, 193–202 (1980). https://doi.org/10.1007/BF00344251
27. Srivastava, N., et al.: Dropout: a simple way to prevent neural networks from overfitting. J. Mach. Learn. Res. **15**(1), 1929–1958 (2014)
28. Zeiler, M.D., Fergus, R.: Visualizing and understanding convolutional networks. In: Fleet, D., Pajdla, T., Schiele, B., Tuytelaars, T. (eds.) ECCV 2014. LNCS, vol. 8689, pp. 818–833. Springer, Cham (2014). https://doi.org/10.1007/978-3-319-10590-1_53
29. Simonyan, K., Zisserman, A.: Very Deep Convolutional Networks for Large-scale Image Recognition (2014). arxiv.org/abs/1409.1556
30. Szegedy, C., et al.: Inception-v4, Inception-ResNet and the Impact of Residual Connections on Learning (2016). arxiv.org/abs/1602.07261
31. He, K., Zhang, X., Ren, S., Sun, J.: Deep residual learning for image recognition. In: IEEE Conference on Computer Vision and Pattern Recognition (CVPR), Las Vegas (2016). https://doi.org/10.1109/CVPR.2016.90
32. Yong-Deok, K., Eunhyeok, P., Sungjoo, Y., Taelim, C., Lu, Y., Dongjun, S.: Compression of Deep Convolutional Neural Networks for Fast and Low Power Mobile Applications (2016). arxiv.org/abs/1511.06530

Autodiagnosis of Information Retrieval on the Web as a Simulation of Selected Processes of Consciousness in the Human Brain

Anna Bryniarska(✉)

Institute of Computer Science, Opole University of Technology,
ul. Proszkowska 76, 45-758 Opole, Poland
a.bryniarska@po.opole.pl

Abstract. An agent searching information on the Web processes finite sets of knowledge descriptions. By analogy to processes of consciousness, the knowledge is (1) assertions – what agent knows, (2) concepts – what agent has knowledge of and (3) axioms – what agent conceives by conceiving rules. An information retrieval system in which the agent knows about itself, is called an autodiagnosis of searching information on the Web. This work presents a theoretical description of autodiagnosis and its references to selected issues of neuropsychology related to processes of consciousness.

Keywords: Autodiagnosis · Neurological models of consciousness
Information model of consciousness · Semantic web
Granular attributive language

1 Introduction

A simulation of consciousness processes in the human brain by the information systems is understood in two ways. First one is understood, accordingly to the bionic defined in 1958 by Jack E. Steele, as a science about systems that behave like a pattern one which is a living system [13]. The second one is understood as an artificial intelligence research on the human brain, which was formulated in 1972 by Arbiib [2]. In these days, mostly analyzed such information systems are artificial neural networks, which are inspired by the biological neural network in the human brain [12]. These neural networks can be, for example, used in the autonomous control systems for cars [4, 16] or other machines (like robots) [17]. The engineering of these systems refers only to such aspects of simulating conscious processes, as representing the searched information in a system environment and regulating the behavior of these systems as their response to the searched information [5]. This approach does not include simulation of such conscious processes that allow us to search information in the environment, react on them and also autodiagnose ourselves based on this information. This autodiagnosis is possible because the human organism can react on the stimuli from the environment. Moreover, the systems, based on this reaction, can autodiagnose

W. P. Hunek and S. Paszkiel (Eds.): BCI 2018, AISC 720, pp. 111–120, 2018.
https://doi.org/10.1007/978-3-319-75025-5_11

themselves based on the searched information [10]. This aspect of information processing by the human brain is connected with a mind [10] and is called *a self-diagnosis of information retrieval* by human. The information retrieval concerns an external and internal environment of human, and also information about himself. In this context, the self-diagnosis is also in some part a self-awareness.

The main motivation of this paper was preparing theoretical project of an information system - the Web agent [6] - which simulates the self-diagnosis of information retrieval. This research is partly inspired by 'The Engineering of Mind' presented by Albus in the paper [1]. As Albus writes 'While the mind remains a mysterious and inaccessible phenomenon, many of the components of mind, such as perception, behavior generation, knowledge representation, value judgment, reason, intention, emotion, memory, imagination, recognition, learning, attention, and intelligence are becoming well defined and amenable to analysis'. That is why the simulation of self-diagnosis is possible. This simulation will be called *the autodiagnosis of information retrieval on the Web*. In this paper is described a model of consciousness and its interpretation in the computer science. Next, there are presented the logical basis of this interpretation, which will be the starting point for the construction of the autodiagnostic algorithm.

2 Neurological Consciousness Model and Its Simulation

In order to develop a scheme of the agent which simulates the autodiagnosis processes of consciousness, are used a neurobiological models of consciousness [9,11,19]. The famous 'binding problem' – the problem of how we mentally represent distinct objects and theirs characteristics – is at least partly resolved by the phenomenon of neuronal synchronization around 40 Hz. But that does not tell us how the representation of a given object enters our consciousness while the representations of others remain unconscious [19].

In the 1990s, Rodolfo Llinas and his colleagues performed a series of detailed studies on thalamocortical interactions [14]. From these studies, they developed a theory that simultaneously integrates the data from the conscious state of waking and from dreaming, addresses the binding problem, and provides a criterion for determining which conscious representation is going to be selected at any given time.

These oscillations are apparently produced by the non-specific nuclei of the thalamus, whose projections pass through the cortex from front to back. The thalamus is often compared to a railroad switching yard, because the signals from all of the senses (except smell) must pass through it before they can reach the cortex. The cortex also sends many connections back to the thalamus. Most of the nuclei in the thalamus are considered 'specific' because their neurons make connections with relatively circumscribed areas in the cortex. Rodolfo Llinas hypothesizes that the oscillations of certain neurons in the thalamus serve as a sort of basic rhythm with which the cortical oscillations of the various sensory modalities synchronize themselves to form a unified image of the environment – somewhat like an orchestra conductor who provides the beat for all the musicians to follow. This is an original solution to the binding problem. In this approach, neurons that are active at the same time are believed to be ‘perceiving the same thing’.

For the described neurological model of consciousness can be designed an information system, simulating processes of consciousness. This system is presented in Table 1.

Table 1. Comparison of the neurological model of consciousness with its simulation.

Neurological model of consciousness [19]	Information system simulating processes of consciousness
The mind: information network in the universe in which human exists [10]	The network of the Semantic Web
The conscious mind, i.e. a mind of specific human: part of the mind, including external and internal environment of this human	The autodiagnosis agent: a program diagnosing actions of the browser and searching on the Web information about realization of this program [6]
The brain - neurological processes	The computer (with the Internet browser)
The cortex: projection and associative areas of cortex	The active browser window: areas in the browser window in which are entered Internet addresses and language expressions
The thalamus	The Internet browser
The cycle of discharging of neurons stimuli on roads cortex-thalamus and thalamus-cortex [13, 19]	The cycle of diagnosing information that is passed from the search engine to the active browser window and other way
The left hemisphere of the brain, in which are correlations of communication by language [19]	The part of browser which shows in the active window searched information as a text in webpages
The right hemisphere of the brain, an interpreter of the language expressions provided by sensory and motor stimuli [19]	The part of browser which shows in the active window searched information as its Internet addresses
The corpus callosum linking both hemispheres and in the cortex area allowing interpret sensory and motor stimuli by language expressions [19]	The interface which allows to search information on the Web or display this information in the active browser window
The process of consciousness: discharging of neurons stimuli in the cortex during the roads cortex-thalamus and thalamus-cortex [19]	The diagnostic process of information displayed in the active window of the browser by autodiagnosis agent, accordingly to the confidence range
Conceiving: The process of consciousness in the associative area of the cortex	Using by the autodiagnosis agent conceiving rules
Consciousness and self-awareness [19]	The cyclical actualization of the diagnose by autodiagnosis agent

3 Information Retrieval on the Semantic Web

The information retrieval systems are widely described in the literature. In this paper the theoretical apparatus of the Information Retrieval Logic *IRL* [6] is widened. The information retrieval logic allows us to search knowledge in the semantic network, especially on the Semantic Web, which is described by the Fuzzy Description Logic (*fuzzyDL*).

In the paper [7] is presented a method of obtaining precise results for searched information, based on the *IRL* logic on the Web and also the postulates of this logic. Furthermore, it is shown that by the defuzzification process can be obtained few results, which are in the confidence range, accepted by experts. The confidence range is a set of acceptable fuzzy degrees of searched information. In this situation, even if we have uncertain, unclear or vague information, we can obtain exact answer about searched information. Then, the positive result of searching information, not only depends on the appropriate choice of the fuzzy degrees acceptable by experts, but also depends on the appropriate description of the concepts and roles interpretation in the appropriate relational structure (a granule system), which corresponds to granulation made by the human brain. The problem of describing such interpretations is formulated in the paper [7].

Furthermore, is used the theoretical apparatus from the paper [8] in order to formulate this problem. To resolve this problem, is proposed to define the granule system and the confidence range as a part of the Web, in which searched information can be considered as credible for experts. For the confidence range, the granule system is a relational structure of the abstract sets.

3.1 The Concept of the Semantic Network

The semantic network of *the first order* is a system:

$$SN = \langle U, AS, \{DS_i\}_{i \in N, i<n+1} \rangle \tag{1}$$

Where sets:

U (individual names) – is a set of unit names of described objects or pronouns indicating singularly on the described objects (identified with *nodes* of the semantic network, corresponds to the Web addresses),

AS (a relational structure) – is a set of the network *edges* designating a certain relational structure.

Descriptive sets $DS_n, n \in N$, are called sets of *the n-argument relations descriptions* and satisfy following conditions:

$$AS \subseteq (DS_1 \times U) \cup (DS_2 \times U^2) \cup ... \cup (DS_n \times U^n) \tag{2}$$

where $ds_i \in DS_i$ is a relations description R such that:

$$\{ds_i\} \times R = (\{ds_i\} \times U^i) \cap AS \tag{3}$$

And the sum of all such relations is equal $U_{gen} = U \times U^2 \times ... \times U^n$.

Let SN^+ be a subsystem of the semantic network SN, considered by experts as the confidence range. The elements of the set AS we called *the assertions.* Any relation R, when it is one-argument relation, which satisfy Eq. 3, is called a concept, while any relation R, when it is two-argument relation, is called a role.

For example, the role 'son' binds a person with name 'John' with a person with name 'Brandon', who is his father. This leads us to assertion: *(son, John, Brandon)*, what can also be written as: *son(John, Brandon)* or *(John,Brandon):son.*

The assertion which is in the sentence 'Eva seats between John and Michael' is written: *seat_between (Eva, John, Michael)* or *(Eva, John, Michael): seat_between.* The roles which are functions of the last argument are called **the operations**, for example in assertion *go_to(John, New York).*

We can notice that in the triple *(Eva, John, Michael)* if we rotate names clockwise, then we get the triple *(John, Michael, Eva)*, which is also an instance of some role. For example this role can be expressed by sentence 'John and Michael seat next to Eva'. This assertion we write as: *(John, Michael, Eva):seat_next_to.* About role *seat_next_to* we say that it is **a cyclical reverse** to the role *seat_between.*

When the triple *(Eva, John, Michael)*, which is an instance of an assertion *seat_between*, we reduce and remove first name, then the pair *(John, Michael)*, is also an instance of some assertion. For example this assertion can be expressed by sentence 'someone seats between John and Michael': *(John, Michael): someone_seat_between.* About this role we say it is **a reduction** of the role *seat_between.*

In the first order semantic network, the set of nodes is extended by concepts and roles of this network as in examples above. Then we get *the second order semantic network.* By extending this network, as long as exists appropriate concepts and roles, we can obtain *the n-th order semantic network.*

3.2 An Information Representation on the Web

In the context of the Semantic Web research [3], the knowledge representation is described by two systems: a terminology called *TBox* and a set of assertion representations called *ABox.*

The Semantic Web can be extended by some nodes which correspond to concepts and roles. It can be also extended by edges which correspond to dependences between these concepts and roles. The descriptions of these dependences are called *axioms*, and the system which represents this knowledge is called *RBox.* In presented research, the knowledge base represented in the attributive language AL of the Description Logic DL, is described as a triple $\langle Ab, Tb, Rb\rangle$, where sets Ab, Tb, Rb are possible to computer processing, finite sets of expressions (descriptions of nodes in this network): assertions, concepts and axioms, respectively. In order to describe approximate knowledge in the information system designated by the semantic network SN (especially the Semantic Web), we use reformulated and expanded AL language. The extension of this language allows us to unambiguously interpret concepts and roles of the Semantic Web in various systems of information granules.

We called this language a granular attributive language (***GAL***; the set of its expressions is denoted as GAL) for the semantic network SN. The part of this language GAL^+, is a granular attributive language for the confidence range SN^+. Let data be some non-empty sets of: individual variables, individual names, concept names, roles names and symbols of concepts modifiers.

Syntax of Concepts and Roles Instances. The concepts instances are symbols $x, y, z, v, ..., x_1, y_1, ...$ of variables and symbols $a, b, c, ..., a_1, b_1, ...$ describing individual names. Variables mean any established individual names. Intuitively, they describe nodes of the semantic network.

The roles instances are tuples $(t_1, t_2, ..., t_k)$ of the concept instances.

For the instance $(t_1, t_2, ..., t_k)$, the transposed instance is $(t_1, t_2, ..., t_k)^T = (t_k, t_{k-1}, ..., t_1)$.

Syntax of TBox. For the set of concepts and roles names are included names:

- $\top$ (Top) – an universal concept or role, includes all instances of concepts and roles,
- $\bot$ (Bottom) – an empty concept or role, is knowledge about lack of any instance of concept or role,
- $\{t\}$ – a singleton of the instance t, concept is uniquely determined by the concept instance t,
- $\{(t_1, t_2, ..., t_k)\}$ – a role which is a singleton of instances tuples.

Let A, B be the concepts names, R is a role name and m is a modifier symbol. Then concepts are:

- $\neg A$ – a concept negation - it means all concept instances which are not instances of the concept A;
- $A \wedge B$ – an intersection of concepts A and B - it means all instances of concepts A and B;
- $A \vee B$ – a union of concepts A and B - it means all instances of concepts A or B;
- $A \backslash B$ – a difference of concepts A and B - it means all instances of the concept A, which are not instances of the concept B;
- $\exists A.R$ – an existential quantification - it means all instances of the concept A which are in the role R with at least once occurrence of the concept A;
- $\forall A.R$ – a universal quantification - it means all occurrences of the concept A which are in the role R with some occurrence of the concept A;
- $m(A)$ – a modification m of the concept A - it means a concept, which is changed concept A by the word m, for example m can have such instances as: yes, no, very, more, most or high, higher, highest; in the approximate calculus modifications are lower or upper approximation.
- $m(R)$ – a modification m of the role R, for $m \in M$,

- R^{-1} – a role cyclically reversed to the role R,
- $A.R$ – a limitation of the role R instances by the instances of the concept A or a property of being the object A in the role R
- R^T – a transposed role of the role R - it means roles that have transposed instances of the role R,
- R^- – a reduction of the role R of a first element - it is a role if R is at least three-argument role and it is a concept if R is at least two-argument role.

Syntax of ABox. For any variables x, y, individual names a, b, a concept name C and a two-argument role name R, assertions are denoted by expressions: $x : C$, $a : C$, $(x, y) : R$, $(a, y) : R$, $(x, b) : R$, $(a, b) : R$. Generally, for the n-argument role R, the assertions expressions are $(t_1, t_2, ..., t_n) : R$, where t_i is any concept instance. The expressions $t_1 : A, (t_1, t_2, ..., t_k) : R$ we read: t_1 is the concept A instance, a tuple $(t_1, t_2, ..., t_k)$ is the role R instance, respectively.

F is a property of the object x, if for some concept A and the role R, $F = A.R$ and $x : A.F$.

Syntax of RBox. For any concepts names A, B, roles names R_1, R_2, and assertion expressions α, β the axioms are expressions:

- $A \subseteq B$ – a conclusion of concepts; $A = B$ – an equality of concepts;
- $R1 \subseteq R2$ – a conclusion of roles; $R1 = R2$ – an equality of roles;
- $\alpha :- \beta$ – a Horn's clause for assertions α, β; we read: if there is an occurrence of the assertion β, then there is also occurrence of the assertion α.

3.3 Semantic of the Attributive Language *GAL*

The expressions of the *GAL* language are interpreted uniquely in different information granules systems [15], defined in: the set theory [3], the stochastic [18], the theory of possible data sets, the fuzzy set theory [7] and the rough sets theory [8]. The *GAL* language expressions are interpreted in chosen granules system:

$$\mathbf{G} =< G, M_G, \cup_G, \cap_G, \setminus_G, `_G, \in_G, \subseteq_G, =_G, 0_G, 1_G, G_0 >, \quad (4)$$

where G is a set of all granules; $\setminus_G$ is a granules difference; $`_G$ is a granules closure; $\in_G$ is a relation of being an element of the granule; $\subseteq_G$ is a relation of the granules conclusion; $=_G$ is a relation of the granules closure. Operations $\cup_G, \cap_G$ are generalized operations of addition and product defined on subsets of the granules G family. For an empty set, value of these operation is an empty granule 0_G and for the set G is a full granule 1_G.

Let a function $\{\}_G : U_{gen} \rightarrow G$ has property $G_0 = \{\{s\}_G : s \in U_{gen}\}$. We define an interpretation $\mathbf{I} = (\mathbf{G},{}^I)$ of the *GAL* language, for which the interpretation function ${}^I : GAL \rightarrow G$ (values ${}^I(E)$ we write as E^I) satisfy following conditions (all interpretations are presented in paper [8]).

I1. For any concept and role instances are assigned granules elements:

$$t^I \in_G x_G, \text{ for some } x \in U, (t_1, t_2, ..., t_k)^I \in_G \{\langle x_1, x_2, ..., x_k\rangle\}_G$$
$$\text{for } \langle x_1, x_2, ..., x_k\rangle \in U^k, t_1^I \in_G \{x_1\}_G, t_2^I \in_G \{x_2\}_G, ..., t_k^I \in_G \{x_k\}_G \quad (5)$$

I2. For concept names C, including singletons $\{t\}$, are assigned granules:

$$\{t\}^I = \{x\}_G, \text{ for some } x \in U, \{(t_1, t_2, ..., t_k)\}^I = \{\langle x_1, x_2, ..., x_k\rangle\}_G$$
$$\text{for } \langle x_1, x_2, ..., x_k\rangle \in U^k, t_1^I \in_G \{x_1\}_G, t_2^I \in_G \{x_2\}_G, ..., t_k^I \in_G \{x_k\}_G$$
$$C^I \in G, C^I = \cup_G\{\{t\}^I : t^I \in_G C^I\} \quad (6)$$

I3. For k-argument role R names are assigned granules:

$$R^I \in G, R^I = \cup_G\{\{(t_1, t_2, ..., t_k)\}^I : (t_1, t_2, ..., t_k)^I \in_G R^I\} \quad (7)$$

I4. For modifiers $m \in M$ are assigned some functions $m^I : G \to G$, where $m^I \in M_G$.

4 Adequate Axioms and Conceiving Rules in the GAL^+

We assume that for the confidence range $SN^+ = \langle U^+, AS^+, \{DS_i\}_{i \in N, i < n+1}\rangle$, the language GAL^+ has interpretations in the granules system:

$$\mathbf{G}^+ = \langle G^+, M_G^+, \cup, \cap, \setminus, `, \in, \subseteq, =, 0_G^+, 1_G^+, G_0\rangle \quad (8)$$

where:

- $G^+ = U_{gen}^+ \cup \wp(U_{gen}^+)$, for $U_{gen}^+ = U^+ \cup (U^+)^2 \cup ... \cup (U^+)^n$;
- $0_G^+ = \emptyset$; $1_G^+ = U_{gen}^+$;
- $G_0 = \{\{s\} : s \in U_{gen}^+\}$, M_G^+ is a set of operations on the set G^+;
- $\cup, \cap, \setminus, `, \in, \subseteq, =$ of standard operations and relations in the set theory.

Based on the set theory, can be defined following highlighted axioms of the GAL^+ language, called adequate, which by any interpretation ^{I} are satisfied in the granules system G^+.

The highlighted adequate axioms are:

Ax1. $\top = \neg\bot, A \subseteq \top, \bot \subseteq A, A \subseteq A$.
Ax2. $A = A, R = R$.
Ax3. $A \vee \bot = A, A \wedge \top = A$.
Ax4. $A \wedge \bot = \bot, A \vee \top = \top$.
Ax5. $A \vee B = B \vee A, A \wedge B = B \wedge A$.
Ax6. $(A \vee B) \vee C = A \vee (B \vee C), (A \wedge B) \wedge C = A \wedge (B \wedge C)$.
Ax7. $(A \vee B) \wedge C = (A \wedge C) \vee (B \wedge C), (A \wedge B) \vee C = (A \vee C) \wedge (B \vee C)$.
Ax8. $A \vee \neg A \subseteq \top, A \wedge \neg A = \bot$.
Ax9. $\neg A \subseteq \top \setminus A$.
Ax10. $\forall C.R \subseteq \exists C.R$.

Ax11. $t : \top :- t : A.$
Ax12. $t : \{t\} :- t : A.$
Ax13. $(t_1, t_2, ..., t_k) : \{(t_1, t_2, ..., t_k)\} :- (t_1, t_2, ..., t_k) : R.$
Ax14. $\exists\{t_1\}.R :- (t_1, t_2, ..., t_k) : R.$
Ax15. $(t_1, t_2, ..., t_k) : R :- \forall\{t_1\}.R.$
Ax16. $(t_2, t_3, ..., t_k, t_1) : R^{-1} :- (t_1, t_2, ..., t_k) : R.$
Ax17. $(t_2, ..., t_k) : R^{-} :- (t_1, t_2, ..., t_k) : R,$ for $k > 2$.
Ax18. $t_2 : R^{-} :- (t_1, t_2) : R.$

The axiom conceiving rule is called an expression $\alpha_1, \alpha_2, ..., \alpha_k/\beta$, for any axioms $\alpha_1, \alpha_2, ..., \alpha_k, \beta$. This rule is adequate for some interpretation function ^{I}, if the interpretation β^I is a result from the interpretations $\alpha_1^I, \alpha_2^I, ..., \alpha_k^I$.

The highlighted adequate conceiving rules are:
Rule1. $A \subseteq B, B \subseteq C/A \subseteq C.$
Rule2. $A \subseteq B, B \subseteq A/A = B.$
Rule3. $A \subseteq B/(t : B) :- (t : A).$
Rule4. $R_1 \subseteq R_2/((t_1, t_2, ..., t_k) : R_2) :- ((t_1, t_2, ..., t_k) : R_1).$
Rule5. $t_k : (\{t_{k-1}\}.(...(\{t_2\}.(\{t_1\}.R)^{-})^{-})...)^{-})^{-}/(t_1, t_2, ..., t_k) : R.$

Adequate conceiving rules are not used for proving theorems. They only establish logical relationships between axioms. If agents in the pragmatic system of knowledge representation use these rules, it means that they correctly conceive axioms interpretations in the granules system.

5 Conclusion

As a simulation of consciousness processes in the left and right hemisphere of human brain, can be proposed autodiagnosis of information retrieval on the Web. This system diagnoses syntax and semantic in the confidence range and also is the autodiagnosis agent which uses conceiving rules as a simulation of consciousness processes. However, there are some problems and tasks which should be considered and solved:

- Prepare autodiagnosis algorithm,
- Use conception of 'conceiving ways' proposed in the paper [8],
- Built logic for information retrieval which theoretical apparatus is presented in this paper,
- Describe autodiagnosis of a concept 'I am' for autodiagnosis agent,
- Consider if frequency of autodiagnostic cycles (frequency of actualizations of autodiagnosis history) should be complies with frequency 40 Hz of the thalamocortical loops,
- Consider if frequency of synchronization of autodiagnosis cycles with autodiagnosis history stability should be complies with brain waves: alpha (8–12 Hz), beta (above 12 Hz), theta (4–8 Hz), and delta (0.5–4 Hz).

References

1. Albus, J.S.: The engineering of mind. Inf. Sci. **117**, 1–18 (1999)
2. Arbib, M.A.: The Metaphorical Brain. An introduction to cybernetics as artifical intelligence and brain theory. Wiley-Interscience, New York (1972)
3. Baader, F., Calvanese, D., McGuinness, D.L., Nardi, D., Patel-Schneider, P.F. (eds.): The Description Logic Handbook: Theory, Implementation and Applications. Cambridge University Press, Cambridge (2003)
4. Behere, S.: Architecting Autonomous Automotive Systems. Licentiate Thesis, Stockholm (2013)
5. Bekey, G.A.: Autonomous robots from biological inspiration to implementation and control. The MIT Press, Cambridge (2005)
6. Bryniarska, A.: An information retrieval agent in web resources. In: IADIS International Conference Intelligent Systems and Agents 2013 Proceedings, pp. 121–125. Prague, Czech Republic (2013)
7. Bryniarska, A.: The paradox of the fuzzy disambiguation in the information retrieval. (IJARAI) Int. J. Adv. Res. Artif. Intell. **2**(9), 55–58 (2013)
8. Bryniarska, A., Bryniarski, E.: Rough search of vague knowledge. In: Wang, G., et al. (eds.) Thriving Rough Sets. SCI, vol. 708, pp. 283–310. Springer, Heidelberg (2017). https://doi.org/10.1007/978-3-319-54966-8_14
9. Cavanna, A.E., et al.: Consciousness: a neurological perspective. Behav. Neurol. **24**(1), 107–116 (2011)
10. Chalmers, D.J.: The Conscious Mind: In Search of a Theory of Conscious Experience. Department of Philosophy, University of California, Santa Cruz (1995)
11. Crick, F., Koch, C.: Towards a neurobiological theory of consciousness. In: Seminars in the Neurosciences, vol. 2. Saunders Scientific Publications, Philadelphia (1990)
12. Friedenberg, J., Siverman, G.: Cognitive Science: An Introduction to the Study of Mind. SAGE Publication, London (2006)
13. Heynert, H.: Einfhrung in die Bionik. VEB Deutscher Verlag der Wissssenschaften, Berlin (1974)
14. Llinás, R., Ribary, U.: Coherent 40-Hz oscillation characterizes dream state in humans. Proc. Natl. Acad. Sci. USA **90**, 2078–2081 (1993)
15. Pedrycz, W.: Granular Computing: Analysis and Design of Intelligent Systems. Taylor & Francis Group, Abingdon (2013)
16. Veres, S.M.: Autonomous and adaptive control of vehicles information - editorial to special issue. Int. J. Adapt. Control Signal Process. **21**(2–3), 93–94 (2007)
17. Veres, S.M., Lincoln, N.K.: Sliding mode control for agents and humans. In: TAROS 2008, Towards Autonomous Robotic Systems, Edinburgh (2008)
18. Walaszek-Babiszewska, A.: Fuzzy knowledge-based approach to diagnosis tasks in stochastic environment. In: Computational Intelligence and Informatics CINTI. IEEE (2013)
19. Young, G.B., Pigott, S.E.: Neurobiological basis of consciousness. Arch. Neurol. **56**(2), 153–157 (1999)

Effect of Spatial Filtering on Object Detection with the SURF Algorithm

Michał Tomaszewski[(✉)], Jakub Osuchowski, and Łukasz Debita

Faculty of Electrical Engineering, Automatic Control and Informatics, Institute of Computer Science, Opole University of Technology, Prószkowska 76, 45-758 Opole, Poland
m.tomaszewski@po.opole.pl, j.osuchowski@doktorant.po.edu.pl

Abstract. The article presents a preliminary study into the detection of electrical insulators in digital images aquired during a power line inspection. Due to the enormous amount of digital data generated during a high voltage lines inspection, there is a need to automate the detection process of power insulators in digital images. As part of the study, the effects of applying spatial filtering into digital images for the purpose of the identification of electrical insulators with the use of a local feature detector and descriptor SURF (Speeded Up Robust Features) were analyzed. The recognition and designation of an insulator's ROI (Region Of Interest) in a digital image will allow the application of more advanced methods aimed at the identification of possible damages in further stages of the analysis.

Keywords: Spatial filtering · SURF algorithm · Object detection
Electrical insulator

1 Introduction

The last few decades have seen a surge in the significance of electrical energy in the economy and society as a whole. With that surge grew the requirements concerning the continuity of power supply. As all technical objects, power lines are subject to the ageing process and failures, which are a natural result of their exploitation. In order to ensure continuous supply of energy, operators of an electrical grid are obliged to make regular preventive inspections, whose main goal is to evaluate the technical condition of power lines. Due to the specificity of transmission infrastructure (vast area, complex structure, exposure to various forms of impact etc.), this is a complicated and time-consuming task requiring expert knowledge and the application of suitable technical measures.

The development of various machine vision technologies for the observation of public space and conservation of various objects, accompanied by a rapid

Michał Tomaszewski, PhD. Eng., Associate Professor; Jakub Osuchowski, Msc. Eng.; Łukasz Debita, Msc. Eng.

W. P. Hunek and S. Paszkiel (Eds.): BCI 2018, AISC 720, pp. 121–140, 2018.
https://doi.org/10.1007/978-3-319-75025-5_12

growth of communication systems, create premises for their broad application in power engineering [1–5] in combination with the computer image analysis technology. Comprehensive recognition of the state of infrastructure objects visible in digital images allows for a broad spectrum of potential applications, such as precise measurement of the position of the structural components of power lines, monitoring that supports emergency population warnings or, indeed, regular maintenance checks and localization of damage to transmission and distribution networks. Controlling the condition of the identified objects with the use of advanced processing methods requires development of solutions dedicated to individual structural components of power lines.

2 Types of Electrical Insulators

An overhead power line is composed of three key elements: electrical conductors, transmission poles and insulators. Overhead power line insulators serve two basic functions. They separate electrical conductors from the ground and the structure of the tower, and they provide those conductors with mechanical support. Overhead power line insulators are built in a way so as to maximize the leakage path from one end of the insulator to the other. This is why they are molded into a series of corrugations or downward-facing cup-shaped surfaces. These "cups" (or skirts) act as umbrellas that keep part of the insulator's surface dry in wet weather [6,7].

Currently, insulators are made out of porcelain, glass or composite materials. Porcelain insulators have been used for over one hundred years and so far they have been appreciated by electrical energy distributors. These are produced with the use of clay, quartz or feldspar, and coated with smooth glaze to shed water more effectively. The quality of these insulators is largely dependent on such parameters as the composition of the porcelain used, production temperature and surface quality [7,8]. Depending on the type of application, porcelain insulators come in several forms. The first form are suspension insulators, which are attached to poles vertically on spindles. The conductors are fixed to the neck of such insulators with the use of special clasps or wires [8–10].

Another type of porcelain insulators are strain insulators. Designed to work in mechanical tension, they are often used with high voltage overhead power lines. They are composed of modules consisting of a disc, cup and pin. The advantage of this type of insulators is that any number of modules can be used depending on the needs. Additionally, in case of damage, one disc can be replaced without the need to replace the whole insulator. Another type of porcelain insulators are stand-off insulators. They consist of a full porcelain cylinder with steel clasps on either end. These are used to support high-voltage lines in substations. Similar to these types of porcelain insulators are long rod line insulators. They are narrower, lighter and used as string insulators. An exemplary photo of a long rod insulator is provided in Fig. 1(a).

Another group of insulators are composite insulators, which currently represent a large part of the market. These insulators offer a multitude of advantages,

This filter outlines the contours that highlight pixels where an intensity variation occurs along the diagonal axes.

Sigma Filter outlines contours by setting pixels to the mean value found in their neighborhood, if their deviation from this value is not important [19]:

$$\text{if } P_{(i,j)} - M > S \text{ then } P_{(i,j)} = P_{(i,j)} \text{ else } P_{(i,j)} = M$$

Given M, the mean value of $P_{(i,j)}$ and its neighbors, and S, their standard deviation, each pixel $P_{(i,j)}$ is set to the mean value M if it falls inside the range $[M - S, M + S]$.

Similarly to Prewitt Filter, **Sobel Filter** emphasizes outer edges of objects. As opposed to the Prewitt filter, the Sobel filter assigns a higher weight to the horizontal and vertical neighbors of the central pixel $P_{(i,j)}$ [19].

$$P_{(i,j)} = max[|P_{(i+1,j-1)} - P_{(i-1,j-1)} + 2P_{(i+1,j)} - 2P_{(i-1,j)} + P_{(i+1,j+1)} - P_{(i-1,j+1)}|], |P_{(i-1,j+1)} - P_{(i-1,j-1)} + P_{(i,j+1)} - P_{(i,j-1)} + P_{(i+1,j+1)} - P_{(i+1,j-1)}|]$$

5 Filters Implemented in MathWorks Matlab Environment

The following spatial filters were implemented in MathWorks Matlab environment:

- Minimal (erosion) filter,
- Maximal (dilation) filter.

One of the basic operations in spatial filtering is **Minimal Filtration**, which narrows the shapes of objects in the processed image based on the defined convolution kernel.

$$P_{(i,j)} = min[P_k]$$

Where $P_{(i,j)}$ denotes the central point of the kernel and P_k is the next pixel depending on the selected shape of the kernel.

Maximal filtering is an operation opposite to erosion, resulting in the expansion of the shapes of objects in the processed image, also based on a defined convolution kernel.

$$P_{(i,j)} = max[P_k]$$

Where $P_{i,j}$ denotes the central point of the kernel and P_k is the next pixel depending on the selected shape of the kernel.

Minimal filtering in MathWorks Matlab environment is implemented by *imerode()* function, while maximal filtering is implemented by *imdilate()* function. The kernels applied for both these kinds of filtering are shown in Fig. 3.

6 SURF and Other Algorithms for Detecting Local Features in Images

The task of detecting matches between two images or objects is part of a number of applications concerned with the methods of processing digital images, i.e. 3D reconstruction, image recording or object recognition. Identifying matches between images may be divided into three main stages. Firstly, algorithms based on the detection of local features detect interest points in given images, such as corners. Next, the neighbors of the keypoints are recorded in the form of vectors which are then matched between different images. Based on their location, vectors of displacement are computed. The quality of the algorithm largely depends on the quality of the determined interest points and their description. Algorithms should detect those points in suitable locations independently of changes to the exposure, location and orientation of the image, and to a certain extent independently of changes to the viewpoint or scale [20].

One of the most commonly used algorithms for the detection and description of local features is SIFT (Scale Invariant Feature Transform). This algorithm transforms an image into a large collection of local feature vectors, generating image pyramids with smaller and smaller resolutions. The images that constitute a pyramid are differentiated, as they are obtained by subtracting two images created as a result of filtering the initial image using the Gaussian filters of different parameters. This operation is therefore a type of low-frequency filtration, which yields gradient images. The pyramids are divided into so-called octaves. The images that make up one octave have the same resolution, but they differ in the scale of the applied Gaussian filter. The subsequent octave contains images with lower resolution, filtered with the use of the same filters. The local features are the minima and maxima detected in the images of the pyramids. A vector for local features (which constitute a type of a histogram of the gradients counted in their neighborhood) is then computed for the designated points. Such local features are mostly insensitive to changes in lighting, rotation and scale thanks to the application of the DoG (Difference of Gaussians) pyramid. Another algorithm characterized by high operation speed and stability of detected keypoints is FAST (Features from Accelerated Segment Test). This algorithm contains a segment test criterion, based on a circle made up of 16 pixels surrounding a candidate point p. The keypoint detector classifies p as a keypoint if the circle contains n contiguous pixels, all of which are brighter than the intensity of the candidate point plus a threshold value t, or darker than $Ip-t$. Due to the fact that the data returned by FAST algorithm cannot be used as a basis for generating a descriptor, SIFT descriptor is often used to describe the features. Another algorithm for the detection and description of features is BRISK (Binary Robust Invariant Scalable Keypoints), which is based ob the FAST spatial-octave algorithm for detection purposes, and on testing binary patterns to describe the detected points. The descriptors for BRISK contain 512 bits. They compute the weighted Gaussian mean on a set derived from a point pattern located in the proximity of a characteristic point. Another algorithm for describing local features is BRIEF (Binary robust independent elementary features). It is a lightweight descriptor which can be easily implemented based on binary number sequences. Binary tests are made based on FERN algorithm, which is the most simplified version of the Bayes classifier. BRIEF descriptor is speed-oriented, which makes it ideal for appliances with lower computational power. Nevertheless, it only describes local features, which first need to be detected. ORB (oriented BRIEF) is an expansion of BRIEF. It incorporates invariance to rotation. It uses a FAST detector, and, as a descriptor, it uses BRIEF that rotates its pattern. Based on strong orientation changes, it selects a suitable pattern [21,22].

For the detection and description of local features, the presented study applied SURF (Speeded-Up Robust Features) algorithm, which is partly based on SIFT, but it is characterized by substantially accelerated speed with only slightly deteriorated accuracy of local feature matching. Another reason for choosing this particular algorithm was that it was already applied, with some success, in the detection of insulators of high-voltage power lines [23]. SURF detection uses the determinant of the Hessian matrix, but it only applies basic approximation methods. The determinant of the Hessian matrix is applied as a measure of local variation around a point, and the point is selected when the determinant is the highest. Given a point $x = (x, y)$ in an image I, the Hessian matrix $H(x, \sigma)$ in x at scale σ is defined as follows:

$$H(x,\sigma) = \begin{pmatrix} L_{xx}(x,\sigma) & L_{xy}(x,\sigma) \\ L_{yx}(x,\sigma) & L_{yy}(x,\sigma) \end{pmatrix}$$

where $L_{xx}(x, \sigma)$ is the convolution of the Gaussian second order derivative $g(\sigma)$ with the image I in point x, and similarly for $L_{xy}(x, \sigma)$ and $L_{yy}(x, \sigma)$.

In order to accelerate computation, the algorithm uses indirect image representation, i.e. the so-called integral image. It is computed based on the input image by summing pixel values. Thanks to the application of this method, the size of the image does not influence the operation of the algorithm. This allows processing of high-resolution images. The entry of an integral image $I_{\Sigma}(x)$ at a location $x = (x, y)$ represents the sum of all pixels in the input image I of a rectangular region formed by the point x and the origin.

$$I_{\Sigma}(x) = \sum_{i=1}^{i \le x} \sum_{j=1}^{j \le y} = I(i,j)$$

SURF descriptor uses Haar wavelet distribution in the region of interest. First, orientation is ascribed to the keypoint. Then, a square region is built around it. This region is then arranged according to the designated orientation for the keypoint. The resulting region is divided into smaller 4×4 regions. Thanks to that, the spatial information is retained. For each of the subregions, features for exemplary points, distributed regularly in the vertices of a 5 × 5 grid, are computed. With the use of Haar wavelet, the dx and dy values are computed. These values are then summed after each of the subregions, and the first set of local features (vector) is generated. The computed values, along with absolute values, create a four-dimension descriptor.

7 Preliminary Assumptions

The comprehensive system for digital image recognition is composed of the following elements:

- low-level processing, which includes image acquisition, initial processing, image quality enhancement (e.g. elimination of noise and other undesirable distortions, increased contrast, filtering etc.),
- medium-level processing, which concerns image segmentation, as well as extraction and description of the local features of objects, (e.g. edge and contour detection, extraction of closed-off regions etc.),
- high-level processing, which involves classification, recognition of objects in digital images and interpretation of the analyzed scene.

The final effects obtained in the last stage depend on the proper recording and preparation of the obtained digital image in preliminary stages. The described works are focused on the effect of the selected filters on the increase in the number of local features present on selected objects in the form of electrical insulators in the analyzed digital images. The applied filters were implemented in National Instruments LabView and MathWorks Matlab environments. The order of individual measures is presented in Fig. 4.

The main objective of the measures taken was to analyze the possibility of enhancing selected elements that match a given pattern (insulator) in an image. Additional objectives include: reduction of noise and distortions in digital

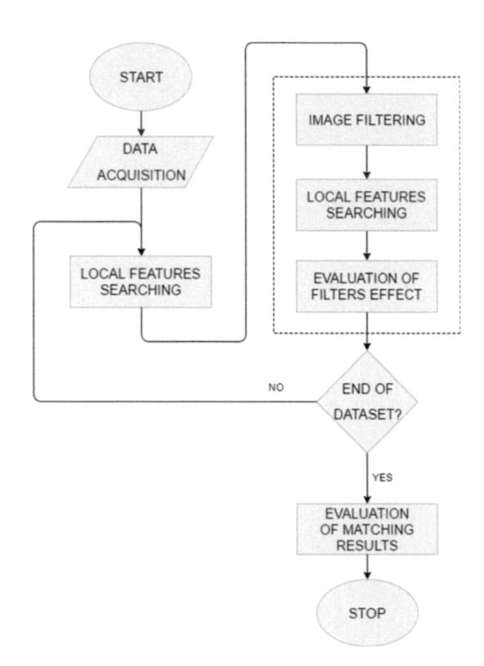

Fig. 4. Stages and course of the study process.

images to the minimum, increase of image quality (sharpening, increased contrast, improvement of low technical quality), and removal of specified image flaws (e.g. blurred objects observed in images recorded by flying and ground vehicles in motion). The set of data on which the study was based was comprised of images obtained with the use of different tools and acquisition methods. The images depicted scenes recorded from a UAV (Unmanned Aerial Vehicle), a helicopter, and as a result of personal inspection by foot. The devices used were Canon 5D Mark II and GoPro Hero 4. The set of data includes scenes with real and working insulators (Fig. 5).

Fig. 5. Selected digital images depicting insulators on existing power lines.

For the purpose of analysis, when detecting an insulator with the use of SURF algorithm, digital images of a porcelain long rod insulator (type LP) were additionally recorded under laboratory conditions. Figure 1 illustrates a scene which was used in further study, and the results of detecting local features with the use of SURF for a non-filtered input image.

Figure 6 illustrates matching of defined local features from the insulator pattern and the local features detected in the image of the laboratory scene.

Fig. 6. Matching of local features between the pattern of an insulator and laboratory scene (scene nr 11): ∘ local feature on the insulator, + local feature outside the insulator.

8 Results of the Analysis

The main criterion in evaluating the effect of different types of filtering on SURF algorithm for the detection and description of local features was the designation of the number of local features detected on the insulator (Fig. 6 ∘) and in the regions of the digital photo outside the insulator (Fig. 6 +). As part of the study, the percentage of the local features that the algorithm detected in the regions containing the insulator was compared with the percentages for the scenes from before filtering, and for different kinds of filtering and different filtering parameters. As a result of the study, we obtained charts of the dependency of the percentage of local features detected on the insulator depending on filtration types and the parameters for different selected scenes (scenes nr 1–10 - photos of insulators on existing power lines, scene nr 11 - photos of an insulator taken in the laboratory). Figure 7 shows the results of the conducted study for the images from the data set used during the analysis.

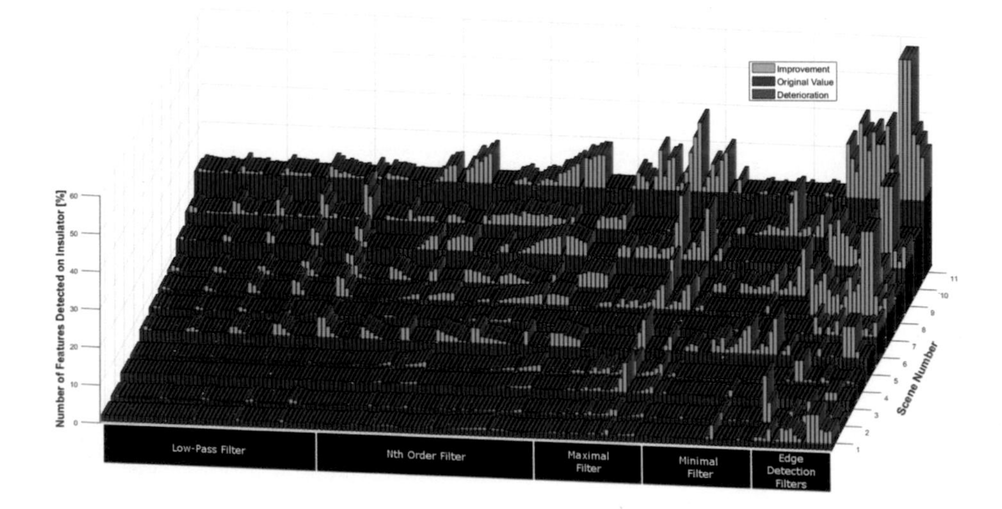

Fig. 7. List of results of the conducted study.

The chart illustrates the effect of filtration on the percentage of the local features detected on the insulator depending on the type of filter and its parameters. Blue color represents the percentage of the local features detected on the insulator without filtration. Green color represents the cases (filter type and parameters) where the percentage of the local features detected on the insulator was increased, and red represents a decrease in the percentage of the local features detected on the insulator.

As can be seen, the applied filters had a noticeable (positive or negative) influence on the number of the detected local features and on their distribution over the surface of the scene (percentage on the insulator). In some cases the percentage of the local features on the insulator was higher by up to 37.19%.

Below are the conclusions derived from the conducted analyses for selected filters and their settings.

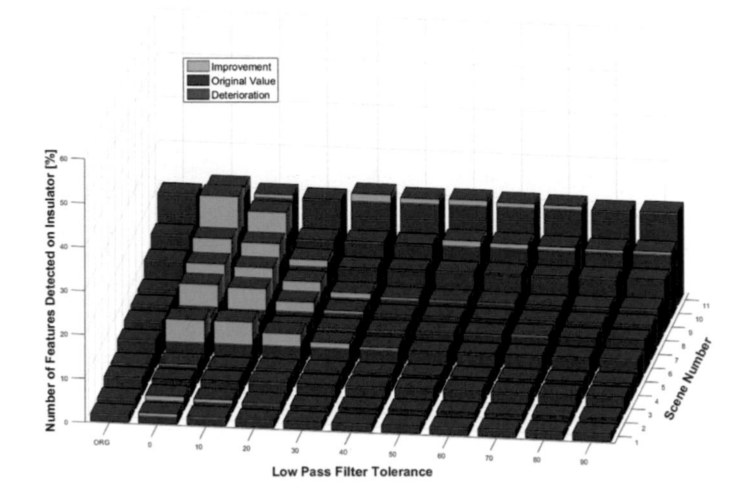

Fig. 8. Dependency between the percentage of the local features detected on the insulator and the parameter setting of a low-frequency tolerance filter with a 7×7 convolution matrix.

Figure 8 illustrates the change in the percentage of the local features detected on the insulator after applying a lowpass filter with a 7×7 convolution matrix. For all the applied sizes of the convolution matrix (3×3, 4×4, 5×5, 6×6, 7×7), the lowpass filter in most of the scenes and types of insulators caused an increase in the percentage of the local features detected on the studied object, especially in the case of a low tolerance value (0, 10, 20). A directly proportional dependency was found between the convolution matrix and an increase in the percentage of the local features detected on the insulator (the larger the matrix, the more local features were detected). Additionally, it was noted that in all of the studied cases (various sizes of the convolution kernels and tolerance values) the lowpass filter did not cause a significant decrease in the percentage of the local features detected.

As can be seen in Fig. 9 the application of the Nth Order Filter had a significant influence on the percentage of the local features detected on the insulator. For some scenes, a change in the order parameter resulted in an increase of the number of the detected local features, and for others that number decreased. In the laboratory scene (no. 11) the negative influence of filtering with the filter order values lower than the median, and a positive influence with the values set to higher than the median, is particularly noticed. Similar results were obtained

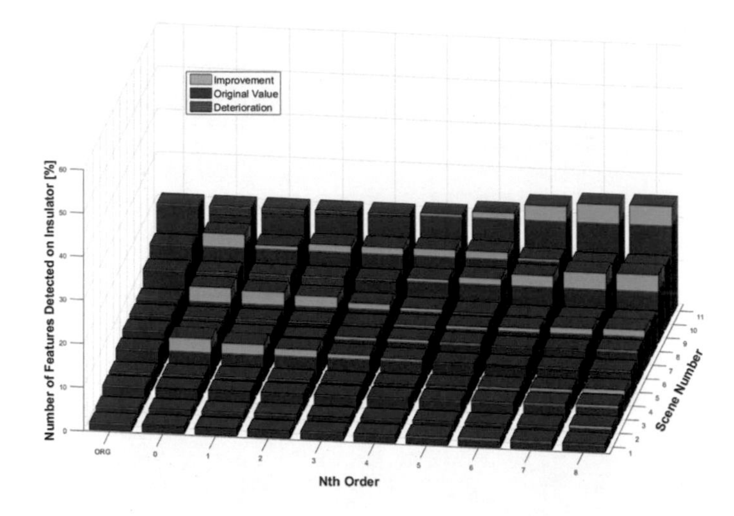

Fig. 9. Dependency between the percentage of the local features detected on the insulator and the order settings of the Nth Order Filter with the convolution matrix of 3×3.

when applying min and max filters and the examples of these results are presented in Figs. 10 and 11.

As for the maximal (dilation) filter, the shape of the convolution matrix had a significant effect on the number of the local features detected on the insulator. An increased percentage of the local features detected on the insulator was observed with the following kernel shapes: Dilate Dimond Radious, Dilate

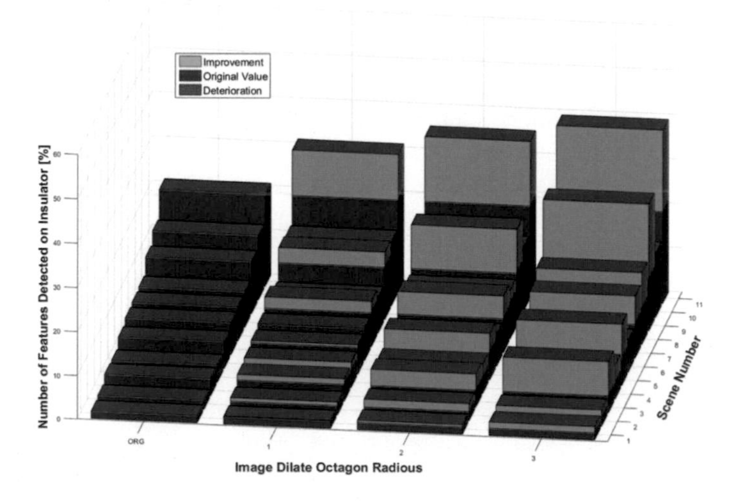

Fig. 10. Dependency between the percentage of the local features detected on the insulator and the radius of the convolution matrix of the dilation (maximal) filter.

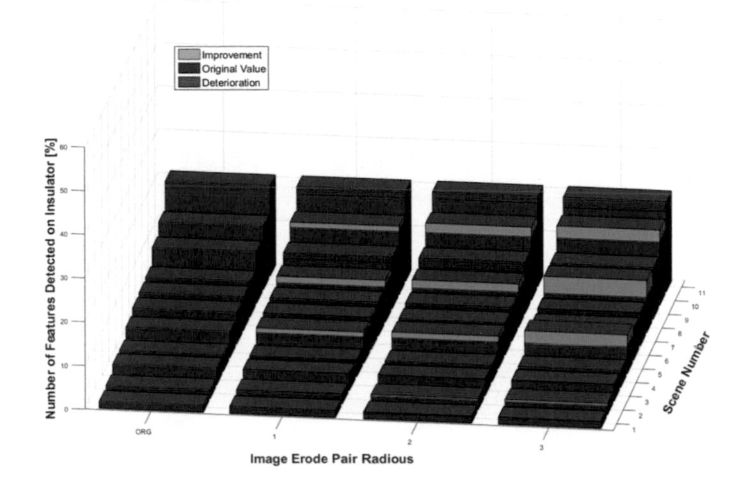

Fig. 11. Dependency between the percentage of the local features detected on the insulator and the radius of the convolution matrix of the erosion (minimal) filter.

Disk Radious, Dilate Pair Radious, Dilate Peridicline Radious, Dilate Octagon Radious, Dilate Line Radious - the highest percentage in the laboratory scene and a smaller (but noticeable) one in the remaining scenes. For the vast majority of the analyzed scenes and different configurations of input parameters, improvement was observed when applying a filter with the peridicline convolution matrix (improvement in 31 out of 33 cases). The most significant positive effect (increase by 18.31%) was observed after applying a filter with an octagon convolution kernel, but the improvement occurred in 27 cases out of 33.

The minimal (erosion) filter had mostly a negative effect on the results of the study (decrease in the number of the local features detected on the insulator). However, an increased percentage of the local features detected on the insulator was also observed, when applying the following convolution kernel shapes: Dilate Peridicline Radious, Dilate Octagon Radious.

The next stage of the study involved an analysis of the effect of the following edge filters: Differentiation filter, Sobel filter, Sigma filter, Prewitt filter, Roberts filter, Gradient filter. Figure 12 illustrates the effect of applying the Sigma filter. Although edge filters increased the percentage of the local features detected on the insulator significantly (green color), they also significantly decreased the overall number of the local features found in the analyzed scene (in one extreme case the number of local features decreased from 2156 in the non-filtered image to 11 local features after applying sigma filtering). As shown in the further stage of the study, such reduction of the amount of local features was insufficient for the detection of the insulator after applying SURF algorithm.

The next criterion of filtering evaluation was an analysis of the matching of local features from the reference image to the local features detected in the output image. The study involved the measurement of the number of the local

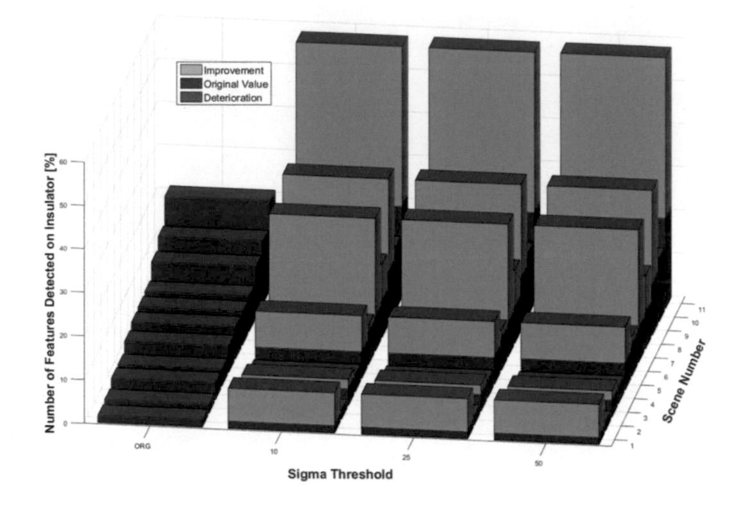

Fig. 12. Dependency between the percentage of the local features detected on the insulator and the threshold parameter of the Sigma filter.

features matched on the insulator before and after geometric transformation, and the duration of detection, description and matching of local features. Moreover, the study also allowed for the way in which the algorithm delineated the ROI. The obtained results were divided into three groups, the examples of which are presented in Fig. 13. The study was carried out only for the laboratory scene (scene 11).

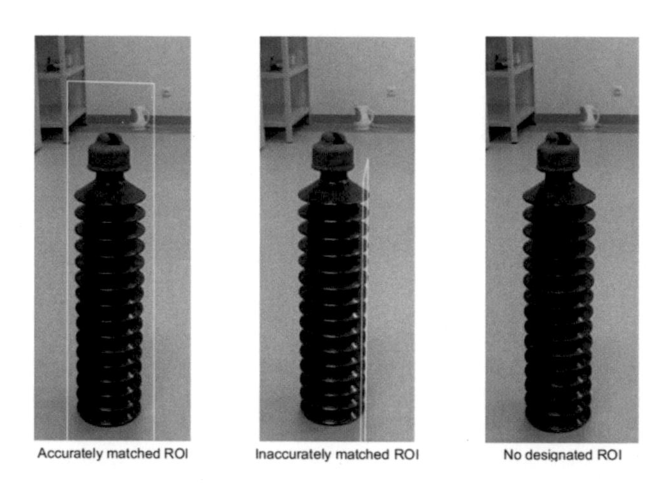

Fig. 13. Various examples of ROI delineation.

Figure 15 illustrates a chart of the dependency between the duration of algorithm implementation (including: duration of local feature detection, duration of descriptor computation and duration of local feature matching between the reference and the analyzed scene) and the number of the matched local features.

In Fig. 15 the green color (rhombus) represents the cases in which the ROI comprises the whole shape of the insulator and distinctly identifies the detected object. Red color (circle) denotes the cases in which the algorithm was unable to determine the ROI at all. Blue color (square) indicates ambiguous cases in which the algorithm did mark the ROI located on the analyzed insulator, but the ROI only comprised a fragment of the searched object, or - in extreme cases - the ROI was determined only as a line located on the searched object. Selected examples of such cases are presented in Fig. 14.

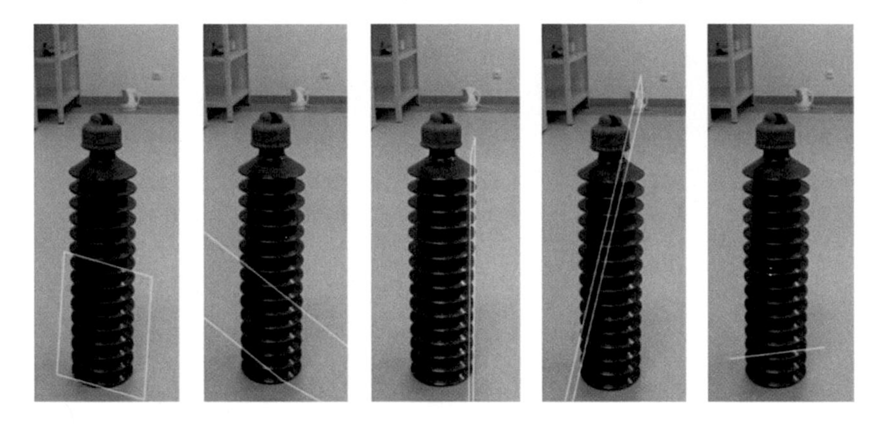

Fig. 14. Selected examples of imprecise ROI delineation.

The minimum duration of accurate ROI marking was 2.5 s. Although shorter durations of algorithm implementation were noted for a smaller number of the local features than the above-mentioned (20), in these cases ROI was not properly marked. Thirty-five local features were identified in the non-filtered image. After filtering, in some cases increased numbers of local features were observed. However, based on the obtained results, it is unclear whether a change in the number of local features affects the accuracy of ROI marking. It was only observed that with a small number of local features (20 descriptors in the analyzed case) the ROI could not be marked properly.

As for edge-detecting filters (in all cases) there was a significant decrease in the number of local features, but in these cases the ROI could not be properly marked. The best results were obtained for the minimal filter. For different filtering parameters (shape and size of the kernel) the algorithm was able to mark the ROI properly (18 cases out of 25–72%). The results for the maximal filter were significantly worse (6 out of 25 cases - 24%). The results of the analysis of ROI marking accuracy in a selected scene in the case of low-frequency and Nth Order filters did not allow us to determine their effect on the success of the studied algorithm for the detection and description of local features.

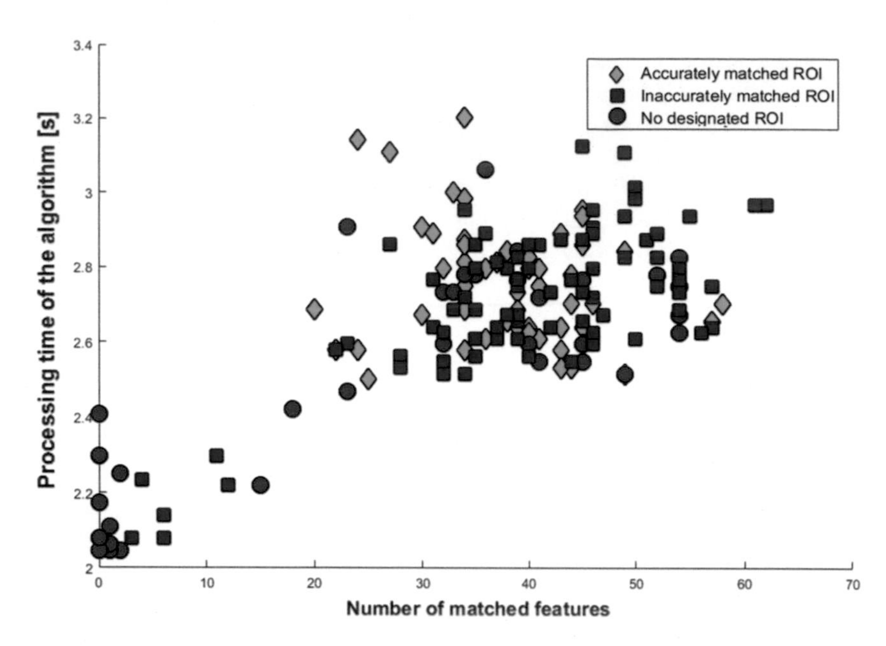

Fig. 15. Dependency between the duration of the algorithm operation and the number of the matched local features.

The above-mentioned results should be treated as strictly indicative, as they require further verification with a significantly greater number of test images.

9 Summary and Conclusions

For the studied scenes, in the case of applying a lowpass filter, in the majority of the analyzed cases the results involved an increased number of local features detected on the insulator. The application of Nth order filters, minimal and maximal, yielded similar results and in some cases the numbers of local features were increased, and in others - decreased. In the case of the laboratory scene, the best results were obtained for a dilation filter (kernel shapes: Dilate Dimond Radious, Dilate Disk Radious, Dilate Pair Radious, Dilate Peridicline Radious, Dilate Octagon Radious, Dilate Line Radious), and for the Nth order filter (order values higher than the median). Edge filters significantly increased the percentage of the local features detected on the insulator, but at the same time they significantly decreased the overall number of the local features found in the analyzed scene, which did not allow the detection of the insulator with the use of SURF.

During the study of the accuracy of ROI marking, carried out for different filters in the laboratory scene, the best results were obtained for the minimal filter. When applying different filtration parameters that affect the number of local features, the algorithm most often marked the ROI properly.

The preliminary analyses presented in this article were carried out based on a relatively small set of data, and a more comprehensive interpretation of the presented mode of action requires a much complex set of data (different types of insulators, various scenes, differentiated equipment and lighting).

The filters applied in the presented analysis were widely available in National Instruments LabVIEW and MathWorks Matlab. A more extensive analysis should involve an expanded set of spatial filters and frequency filters. Further studies are to implement the whole study process in one programming environment, which will facilitate the process of experimentation and allow for the analysis of the durations of all computation stages.

References

1. Zator, S., Michalski, P., Lasar, M.: Linking chromatic images with spatial data. In: PAK, vol. 11, pp. 962–964 (2012)
2. Zator, S., Michalski, P.: Pomiar zmian topologii obiektów na podstawie wyznaczania parametrów położenia markerów. In: PAK, vol. 12, pp. 1480–1482 (2011)
3. Tomaszewski, M., Krawiec, M.: Detection of linear objects based on computer vision and Hough transform. Przeglad Elektrotechniczny (Electrical Review), vol. 88 10b (2012)
4. Osuchowski, J.: Insulator detection based on hog features descriptor. Przeglad Naukowo-Metodyczny Edukacja dla Bezpieczeństwa, vol. 34, pp. 1176–1186 (2017)
5. Zator, S., Gasz, R.: A analysis of the geometry HV power line uploaded with used high resolution images. In: PPM, pp. 127–130 (2014)
6. Holtzhausen, J.: High voltage insulators. Electr. Eng. **53**(11) (1994)
7. Anjum, S.: A Study of the Detection of Defects in Ceramic Insulators Based on Radio Frequency Signatures. University of Waterloo (2014)
8. Cotton, H., Barber, H.: The Transmission and Distribution of Electrical Energy Paperback. Hodder Arnold, London (1970)
9. Han, S., Hao, R., Lee, J.: Inspection of insulators on high-voltage power transmission lines. IEEE Trans. Power Del. **24**(4), 2319–2327 (2009)
10. Online Electrical Engineering Study Site. http://www.electrical4u.com/types-of-electrical-insulator-overhead-insulator
11. Lee, J., Park, J., Cho, B., Oh, K.: Development of inspection tool for live-line insulator strings in 154kV power transmission lines keywords power transmission lines keywords. In: Power, vol. 7, no. 1 (2007)
12. Oberweger, M., Wendel, A., Bischof, H.: Visual recognition and fault detection for power line insulators. In: Computer Vision Winter Workshop (2014)
13. Wang, X., Zhang, Y.: Insulator identification from aerial images using support vector machine with background suppression. In: ICUAS (2016)
14. Chojnacki, A.: Analysis of reliability of selected devices in MV/LV substations (2011)
15. Bretuj, W., Fleszy, J., Wieczorek, K.: Diagnostyka izolatorów kompozytowych eksploatowanych w liniach elektroenergetycznych. Przeglad Elektrotechniczny (Electrical Review), vol. 5a, no. 5 (2012)
16. INMR World Congress. http://www.inmr.com/examples-insulator-failure
17. INMR World Congress. http://www.inmr.com/overview-failure-modes-porcelain-toughened-glass-composite-insulators/2

18. Tadeusiewicz, R., Korohoda, P.: Komputerowa analiza i przetwarzanie obrazów. Wydawnictwo Fundacji Postępu Telekomunikacji, Kraków (1997)
19. IMAQ Image Processing Manual, National Instruments Corporation (1999)
20. Bay, H., Tuytelaars, T., Van Gool, L.: SURF: Speeded Up Robust Features. In: ECCV, vol. 3951 (2006)
21. Mikolajczyk, K., Schmid, C.: Indexing based on scale invariant interest points. In: ICCV, vol. 1, pp. 525–531 (2001)
22. Lowe, D.G.: Object recognition from local scale-invariant features. In: ICCV, vol. 2 (1999)
23. Zhao, Z., Liu, N.: The recognition and localization of insulators adopting SURF and IFS based on correlation coefficient. Optik **125**, 6049–6052 (2014)

Concept of Brain-Controlled Exoskeleton Based on Motion Tracking and EEG Signals Analysis

Andrzej Olczak(✉)

Faculty of Electrical Engineering, Automatic Control and Informatics,
Institute of Computer Science, Opole University of Technology,
Prószkowska 76, 45-758 Opole, Poland
itandrzejolczak@gmail.com

Abstract. The article describes the possibilities of motion tracking and electroencephalography as the methods through creating a new type of powered exoskeleton control system. Modern motion tracking methods for three-dimensional space were presented with their advantages and disadvantages. Brain-computer interface was introduced as a possible control system for the robotic exoskeleton. Combined data model was proposed as the hypothetical solution based on electroencephalographic signals for the steering method.

Keywords: Motion tracking · Brain-computer interface
Powered exoskeleton · Control system · Electroencephalography
Electromyography

1 Introduction

Motion tracking is a process of acquisition the movement data measurements for each three axes in three-dimensional Euclidean space, in the function of time. In the last few years motion tracking methods have reached a considerable development. Those methods have a multiple professional applications, beginning with game industry, going through surface measurements, navigation systems, military, sports motion analysis, object selection or manipulation and ending with medical area, especially in surgery applications [1].

Brain-computer interface (BCI) is the way of communication between brain and computer. It collects the data of neural activity, which is generated by brain and transfer it to the computer for further transformations. Normal output pathways of muscles and peripheral nerves are excluded from the signal [2].

Powered exoskeleton is the artificial mechanical skeleton on the outside of the body, in which kinematics are commonly implemented with servomechanisms controlled by the central driver. Exoskeletons are often used by physically handicapped disabled people to restore and regain the physical movements. They are also used in military to increase the strength of soldiers [3].

The concept assume parsing the acquired motion tracking data and the neural activity data from EEG to create the control system for the powered exoskeleton.

W. P. Hunek and S. Paszkiel (Eds.): BCI 2018, AISC 720, pp. 141–149, 2018.
https://doi.org/10.1007/978-3-319-75025-5_13

This method can also be used for a many more applications, such as steering humanoid robots with wireless interfaces, steering robotic arm, etc.

2 Motion Tracking Methods

Nowadays, there are a plenty of motion tracking solutions. The oldest methods, based on electro-mechanical systems, were innovated for the game industry. The most popular solution was the joystick, which processed the movement into simple two-dimensional direction data and usually has some buttons to increase the possibilities. Another popular solutions were pads and old generation of computer mouse devices. The amazing technological complex solution based on that system type is the da Vinci Surgical System. It is a real-time surgical medical robotic system, where doctor operate by triggers and grippers on the user console, the view is shown on the screen, the movement is repeated by the robotic arms on the patient. In the case of the exoskeleton, it can have by oneself integrated electro-mechanical parts for movement and angles measuring.

Motion tracking based on microelectromechanical systems (MEMS) is the next method. Accelerometer, gyroscope and magnetometer are the sensors which can be used to track the movement. Accelerometer measures the current acceleration, gyroscope - the rotation velocity and the magnetometer - the magnetic field in the surround of device. All of those sensors operate on three-dimensional coordinate system. Because of that the acceleration is the second derivative of a way with respect to time, the movement can be estimated within physical equations. On the accelerometer sensor indications, the force due to gravity has a strong influence to an acceleration values. Therefore, the gravitational acceleration values must be eliminated from the signal to get the proprietary acceleration of the device. Furthermore, the zero-g offset has an impact on this signal. It is a typical noise source of the sensor, which results from device production imperfection. Noise filtering process, based only on the accelerometer data is heavy to accomplish and not always gives efficient results, thus this data is combined with the data from gyroscope and magnetometer signals and this solutions is termed “sensor fusion” method. Gyroscope sensor have one fundamental flaw, which is the drift and means that the angle values are deviating in a longer time interval. The magnetometer signal is helpful to get the constant waypoint of direction, because this sensor lead the position of Earth’s magnetic field. Nevertheless, the other external magnetic fields biases depending on the magnetometer can disrupt the motion measurements [4]. The most interesting innovation in that type of technology was made by Xsens Technologies B.V. as the product named Xsens MVN, which is the wearable suit with embedded motion sensors. Movement output data is processed with the sensor fusion algorithms into 3D kinematic model using biomechanical model [5]. This solution is commonly used in the cinematography as a motion capture method for the animated movie characters [6].

Another motion tracking method is based on the point cloud technology. The main principle of this technology depends on building three-dimensional

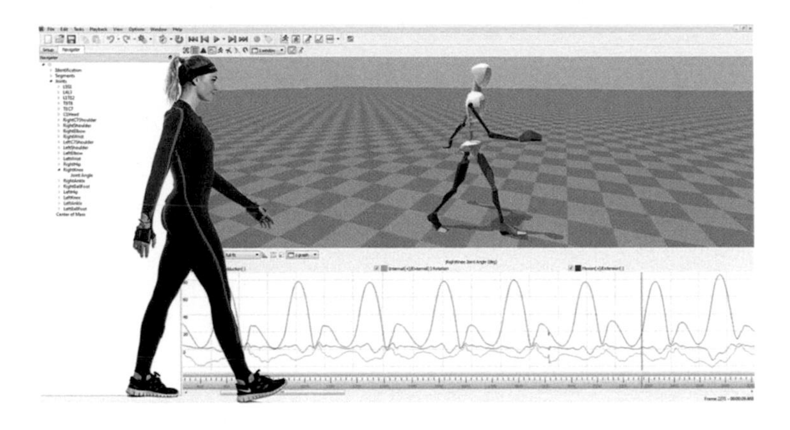

Fig. 1. Motion measurements with XSens MVN [7]

objects from the real world acquired by the external devices, such as cameras, 3D scanners, tomographs or even radars (Fig. 1).

Point cloud methods based on camera devices are performed most often with two popular methods. First is implemented among others within Microsoft Kinect. The functional principle of this method consists of emitting the grid of markers as infrared light points within the IR projector and receiving this projection with IR camera. The application of infrared light allows to hide the visibility of projected points, because the frequency of those light is invisible for human eye. The depth of the objects in scene is estimated with an internal triangulation process of projected scene with infrared lights and objects, where lights are shifted. With an image correlation process a disparity map is calculated. The distances to the sensor are retrieved from the corresponding disparity pixels [8]. Kinect have second RGB camera, but it is used for other features, such as webcam, textures mapping, face recognition, etc. This technology have some limitations, like the infrared light interference on direct sunlight exposure, objects recognition reduced to the distance between approximately 0,5 m to 3 m [9]. Device construction is shown in Fig. 2. Second motion tracking method based on point cloud is implemented with Leap Motion controller. This device was innovated to simplify the usage of three-dimensional user interfaces by a hands gestures recognition. It has built-in two cameras and three infrared LEDs, which track light in wide angle lenses for a large interaction space, where the shapes are taken of an inverted pyramid, which represents the intersection of the binocular cameras fields of view. The image data is streamed to the computer via USB and is retrieved by image processing software to reconstruct a 3D representation of device view, without generating a depth map. Due to limitation of LED light propagation through space, maximum distance for this method equals 80 cm [10]. Point cloud methods based on image processing require efficient computers, because the appropriate algorithms performs high amount of calculations.

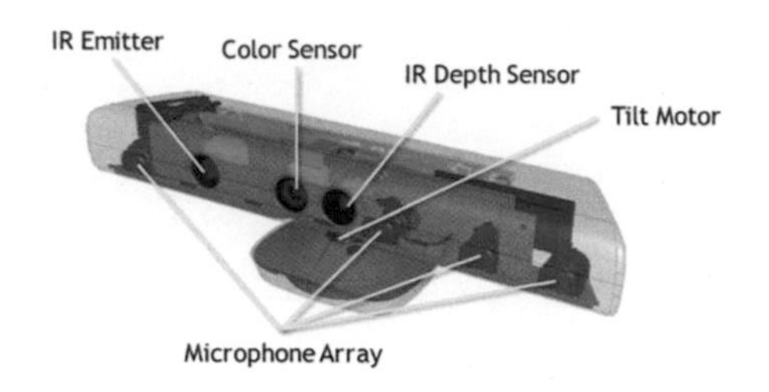

Fig. 2. Microsoft Kinect construction [11]

In the case of movement data capturing, the most accurate results and high computing efficiency gives the radar technology. The method is based on electromagnetic waves emitting in a broad beam. Waves are scattered by real objects and some energy portions are reflected back towards the radar antenna. Object characteristics and dynamics are computated from signal parameters, such as energy quantity, frequency and time delay. On the grounds of gained signal it is possible to build three-dimensional dynamic object with full movement characteristics. This solution is used in the project Soli provided by Google company [12]. The affirmation of this technology importance is that, it has been used in all Tesla cars as a supplementary sensor to increase the performance of standard camera image processing system based on point cloud, essentially when the weather is foggy, dusty, rainy and snowy, therefore when RGB camera have a lot of noise in signal. The main weakness of that technology is that some materials are transparent for radars [13].

3 Electroencephalography

Electroencephalography (EEG) is a method of reading, recording and analyzing electrical potentials generated by the central nervous system (CNS) within the electroencephalogram device, which are in the weak order of 5–100 μV. It is the most popular method of brain-computer interface (BCI) systems. Signal acquisition is performed by the electrodes placed on the human scalp within the conductive paste to decrease a contact impedance, therefore it is noninvasive communication channel. Difference between the potential of the signal electrode and the reference electrode is an output value. The determined frequency bands corresponding with a specific brain rhythms are examined. Those bands are:

- delta (0.5–4 Hz), human is in a deep dream,
- theta (4–7 Hz), drowsiness, early slow-wave sleep,
- alpha (8–13 Hz), relaxed wakefulness,
- beta (13–30 Hz), more prominent mental activity,
- gamma (40–100 Hz), human in movement.

After acquiring signals generated by cortical activity, brain-computer interface perform some preprocessing due to high levels of signals noise and interference.

In the next step, features which corresponds to specific EEG components are estimated and classified. For an effective communication user should participate in a couple of training sessions [14]. Extracted and classified features can represent accurate reflection of human mental activity, including motor control activity. The sample of this device type is presented in Fig. 3.

Fig. 3. Emotiv EPOC+ headset [15]

4 Electromyography

Electromyography (EMG) is a method of acquiring, recording and analyzing bioelectrical signals, defined as myoelectric signals, generated by muscles and read within the EMG sensor and electrodes. Those signals reflects the neural-muscles activity and its respond to the behavior tasks, functional movement, work, training or therapy, therefore it is a complex muscle functionality examination. The most popular electrode types are superficial and needle electrodes, where only the first one belong to noninvasive method. The realization of exploration with this method is to put two electrodes on a particular muscle with 1–2 cm distance between them. Those electrodes are used for a signal sensing. Besides, one reference electrode (termed also ground electrode) must be also set, on an electrically

Fig. 4. Robotic arm control via EMG signal [17]

neutral tissue. It is used for a quality and signal adequateness evaluation. In the next step, raw signal is processed in order to filter the disruptions, which can be inherent noise in the electronics components, ambient noise of electromagnetic radiation, motion artifacts from the interference between detection surface and the skin, inherent instability of the signal and EKG artifacts. Electromyography is directly used to estimate the force generated by the muscle, to determine the activation of the muscle and to analyze the muscle index of the rate. This technique applies in medical experiments, rehabilitation, ergonomics, sports science [16]. EMG-based control of a robot arm is shown in Fig. 4.

5 Data Fusion Model

In order to create a powered exoskeleton using brain-computer interface as a steering method it is important to combine the data from motion tracking methods and the data from BCI. Data fusion is the process of combining multiple data sources into one, consistent and accurate integrated data source.

Motion tracking methods can generate movement data distances with respect to time, regardless of used technique. The best choice to create the powered exoskeleton control system will be to use the XSens MVN solution, because it can map the movements directly into three-dimensional data set and is highly resistant for a noise sources. Owing to the fact that this solution is expensive, using the Microsoft Kinect for a testing platform is also good enough, because it can map a whole body, but with lesser accuracy.

Within the electroencephalography method the signal shows changes in mental activity as an electrical potentials and is recorded with respect to time. Due to this method it is possible to extract the motor control activity data from a whole signal. With regard to the signal noise and sometimes user frustration in EEG experiments, which can change his mental state, the correct signal processing can be improved with an additional electromyography analysis in the same time as a verification method [14].

Data model based on combining data sets from motion tracking measurements and electroencephalography method correlated in a function of time can be a groundwork for knowledge how human brain reacts on exact, direct body movements performed in a three-dimensional space.

6 Powered Exoskeleton Control System

Nowadays, there are several methods for control the powered exoskeleton. The most popular are the force and torque verification at real time, techniques based only on the force sensors and methods performed within the electromyography signals processing [15]. Example of this type of device is presented in Fig. 5. Powered exoskeletons controlled by brain-computer interfaces were also innovated. These interfaces belong to a group termed brain-machine interfaces (BMI), technical solutions which translates neuronal activity information into commands capable of controlling external devices or software. One of the current methods concentrate on exact things, what can be recognized with EEG signal analysis after some trainings. Solution based on that method was showed with spectacular public appearance on the World Cup football competition when the first ball of the tournament was kicked by a paralyzed person wear in robotic suit [18]. Another method is about observing five LEDs, which flickers at a different frequency. When the person focusses attention to one specific diode, reliant frequency band is reflected within the EEG readout, therefore exact signal is used to control the powered exoskeleton [19].

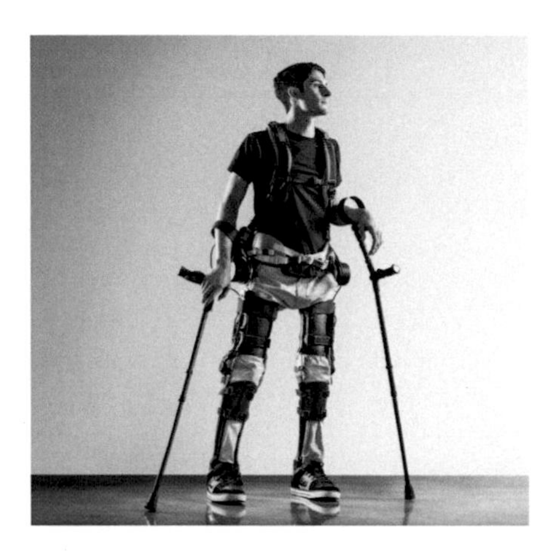

Fig. 5. Phoenix robotic exoskeleton [20]

Because of the fact that BMI solutions described above are not intuitive methods, developing a new method based on natural user thoughts is worth considering. A good hypothetical choice seems to be the possibility of using the motion tracking and EEG data fusion model. Process, where the brain activity signal was recorded while the movements with measurements where done, can be inverted. In that case, the mental activity which was before recorded, can be a basis for a similar or this same activity recognized once again to perform the control of the exoskeleton robotic limbs, where original movements are mapped into exoskeleton movements.

7 Summary and Conclusions

Powered exoskeleton based on described concept can be a great solution for handicapped people to enhance and restore functional ability and quality of life within rehabilitation trainings and sessions. It can be applied to substitute the amputated limbs or even can be used when spinal cord was injured. Modern solutions are not giving satisfying results because of complicated control methods. System based on the mental potentials can comprise this remedy as the natural user interface control method.

Contemporary development of motion tracking methods allows to get the correct movement measurements for the data model. By a contrast, an electroencephalography signal is strongly disturbed by a noise. In any case, signal related to the motion must be dissected from whole signal band. The first task as well as the second is very difficult to accomplish. Many artifacts removal methods from EEG signal was developed and can be used to improve the signal processing. Moreover, for the recording brain activity when human is in movement the EMG method can be applied as a verification solution. In the case of thinking about real control system for the powered exoskeleton, all impediments must be removed. In a test environment, steering can be applied with 3D visualization on the computer.

Important issue is to verify the correctness of this solution on greater statistical assay. Other essential thing is to improve the possibilities of exoskeleton by adding the feedback as the response for a collisions. A sound, vibration or information on the screen can be used as the returnable response signal.

References

1. Welch, G., Foxlin, E.: Motion tracking: no silver bullet, but a respectable arsenal. IEEE Comput. Graph. Appl. **22**(6), 24–38 (2002)
2. Vallabhaneni, A., Wang, T., He, B.: Brain-Computer Interface. Neural Engineering, Bioelectric Engineering, pp. 85–121. Springer, Boston (2005)
3. Hong, Y.W., King, Y., Yeo, W., Ting, C., Chuah, Y., Lee, J., Chok, E.T.: Lower extremity exoskeleton: review and challenges surrounding the technology and its role in rehabilitation of lower limbs. Aust. J. Basic Appl. Sci. **7**(7), 520–524 (2013)
4. Schall, G., Wagner, D., Reitmayr, G., Taichmann, E., Wieser, M., Schmalstieg, D., Hofmann-Wellenhof, B.: Global pose estimation using multi-sensor fusion for outdoor augmented reality. In: 8th IEEE International Symposium on Mixed and Augmented Reality, Orlando, FL, pp. 153–162 (2009)
5. Roetenberg, D., Luinge, H., Slycke, P.: Xsens MVN: full 6DOF human motion tracking using miniature inertial sensors, Xsens Motion Technol. BV Technical report, vol. 3 (2009)
6. Online XSens Technologies B.V. Site (2017). https://www.xsens.com/productions-powered-xsens
7. Online XSens Technologies B.V. Site (2017). https://www.xsens.com/wp-content/uploads/2014/11/MVN-BIOMECH-Link.jpg
8. Alhwarin, F., Ferrein, A., Scholl, I.: IR stereo kinect: improving depth images by combining structured light with IR Stereo. In: PRICAI 2014: Trends in Artificial Intelligence: 13th Pacific Rim International Conference on Artificial Intelligence, Gold Coast, QLD, Australia, pp. 409–421. Springer (2014)
9. Maimone, A., Fuchs, H.: Encumbrance-free tele-presence system with real-time 3D capture and display using commodity depth cameras. In: 10th IEEE International Symposium on Mixed and Augmented Reality, ISMAR 2011, pp. 137–146 (2011)
10. Online Leap Motion Inc., Site, Alex Colgan, How Does the Leap Motion Controller Work (2017). http://blog.leapmotion.com/hardware-to-software-how-does-the-leap-motion-controller-work
11. Online Microsoft Developer Network Site, Kinect for Windows Sensor Components and Specifications (2017). https://msdn.microsoft.com/en-us/library/jj131033.aspx

12. Online Google Project Soli Site (2017). https://atap.google.com/soli
13. Online Tesla Inc., Web Site, Upgrading Autopilot: Seeing the World in Radar (2016). https://www.tesla.com/en_EU/blog/upgrading-autopilot-seeing-world-radar?redirect=no
14. Ebrahimi, T., Vesin, J.-M., Garcia-Molina, G.: Brain-computer interface in multimedia communication. IEEE Sig. Process. Mag. **20**, 14–24 (2003)
15. Online EMOTIV Inc., Site (2017). https://www.emotiv.com/product/emotiv-epoc-14-channel-mobile-eeg/
16. Konrad P.: ABC EMG – Praktyczne wprowadzenie do elektromiografii kinezjologicznej, Technomex Sp. z o.o. (2007). ISBN 83-920818-1-1
17. Castellini, C., Hornung, R., Vogel, J., Urbanek, H.: Bio-data exploitation for supportive robotics, Online Deutsches Zentrum für Luft- und Raumfahrt Site, Institute of Robotics and Mechatronics (2017). http://www.dlr.de/rm/en/desktopdefault.aspx/tabid-9277/15984_read-39321/
18. Online LiveScience Site, Tanya Lewis, Staff Writer: World Cup Exoskeleton Demo: Hope or Hype? (2014). https://www.livescience.com/46285-world-cup-exoskeleton-demo.html
19. Kwak, N.-S., Müller, K.-R., Lee, S.-W.: A lower limb exoskeleton control system based on steady state visual evoked potentials. J. Neural Eng. **12**(5), 056009 (2015)
20. Brewster, S.: This $40,000 Robotic Exoskeleton Lets the Paralyzed Walk, Online MIT Technology Review Site (2016). https://www.technologyreview.com/s/546276/this-40000-robotic-exoskeleton-lets-the-paralyzed-walk/

System to Communicate Disabled People with Environment Using Brain-Computer Interfaces

Natalia Browarska(✉) and Tomasz Stach

Faculty of Electrical Engineering, Automatic Control and Informatics,
Opole University of Technology,
Proszkowska 76, 45-271 Opole, Poland
natalia.browarska@gmail.com, stach.tomasz@gmail.com

Abstract. In this chapter a project of a system to communicate disabled people with environment using Brain-Computer Interfaces was described. The user interface was developed basing on a pictogram writing system. In the test phase EPOC+ Neuroheadset by Emotiv was used. The system was adapted for a young disabled girl with Dandy-Walker syndrome (DWS). The girl uses this type of communication with her family under the care of a speech therapist on a daily basis.

Keywords: BCI · EEG signal · Pictogram writing system
Disabled people

1 Introduction

The human brain is considered to be the most complicated computer in the world. So far no one has successfully managed to simulate the entire brain work. Thanks to development of medicine and information technology electroencephalography examination (EEG) was used to build Brain-Computer Interfaces (BCI) technology. BCI devices are mainly dedicated for disabled people e.g. to control electrical wheelchairs, neuroprostheses, exoskeletons, speech prostheses, intelligent home systems or any computer applications [5]. In this chapter a concept of using BCI to handle a dedicated communicator for disabled people based on a pictogram writing system was described.

2 Brain-Computer Interfaces

Every year Brain-Computer Technology is applied for new tasks. Depending on the user's needs, BCI can be used in different areas of life. Brain-Computer Interfaces are successors of nowadays mainly used mechanical interfaces such as: a computer mouse, a keyboard or a joystick - controlled by human muscles. BCI let people communicate with external devices. This correlation allows wireless

W. P. Hunek and S. Paszkiel (Eds.): BCI 2018, AISC 720, pp. 150–157, 2018.
https://doi.org/10.1007/978-3-319-75025-5_14

communication (without using muscles) [5]. BCI technology combines many scientific disciplines, such as: biomedical engineering, IT, medicine, electronic and signal analysis, automatic control and robotics [4,10]. The main idea of using BCI is to help disabled people handle a computer, a wheelchair or neuroprostheses. BCI devices can be also applied in entertainment - neurogaming [11]. With the Emotiv device people are able to compose music using only one instrument - their mind. Mindtunes is a project that was developed through the collaboration of Smirnoff, Dj Fresh and a neurotechnology expert - Julien Castet. Thanks to the possibility of detecting emotional states of users and electrical neuroimaging, BCI was implemented in a marketing research - neuromarketing [7]. Additionally, marketers are able to integrate EEG signals with eye tracking data. The results are more satisfying than in a standard marketing research [6]. For a daily use BCI devices can be paired with smartphones e.g. to monitor biological functions when working out or when sleeping. There are two leading BCI companies which develop this kind of devices: Emotiv Inc. and Neurosky.

3 Emotiv EPOC+ Neuroheadset

The device made by Emotiv Inc. is the most recognized commercial device. EPOC+ Neuroheadset (shown in Fig. 1) is a 14 channel wireless EEG. The device uses sensors to identify electrical signals generated by human brain in order to detect user's thoughts, feelings and facial expression. All this just by wireless connection with a computer. This device has also a build-in gyroscope [1–3].

Fig. 1. Emotiv EPOC+ Neuroheadset device.

The software shipped by Emotiv Inc. allows extraction of a raw EEG signal, but the most popular tool among users of EPOC+ Neuroheadset is EPOC Control Panel. In this software the user can communicate nonverbally with a computer by EPOC+ Neuroheadset. The user can assign some action or a key press to specific types of a facial expression. Working with this software is much easier due to the animated avatar - Emobot (Fig. 2) which imitates the user's facial expression.

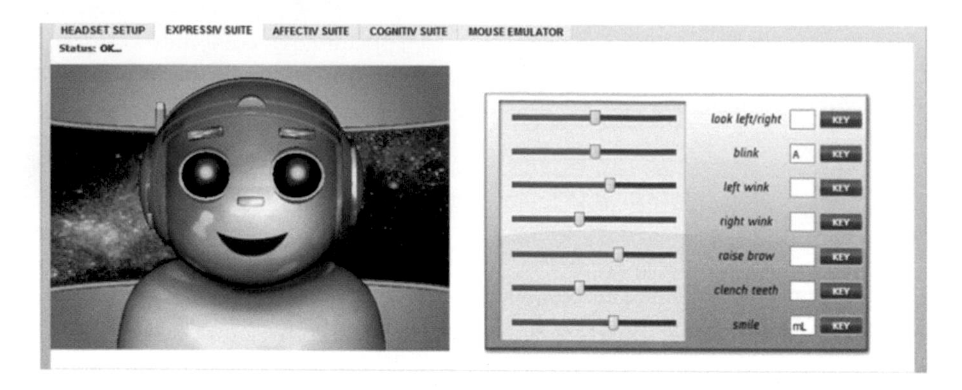

Fig. 2. Emobot avatar in EPOC control panel.

4 Concept of Brain-Computer Interfaces

The aim of Brain-Computer Technology is EEG examination - a non-invasive diagnostic method which records electrical activity of the brain [9]. Electrodes are placed along the scalp. This examination measures voltage of potential changes in particular parts of the brain. An example of the electroencephalogy record is shown in Fig. 3.

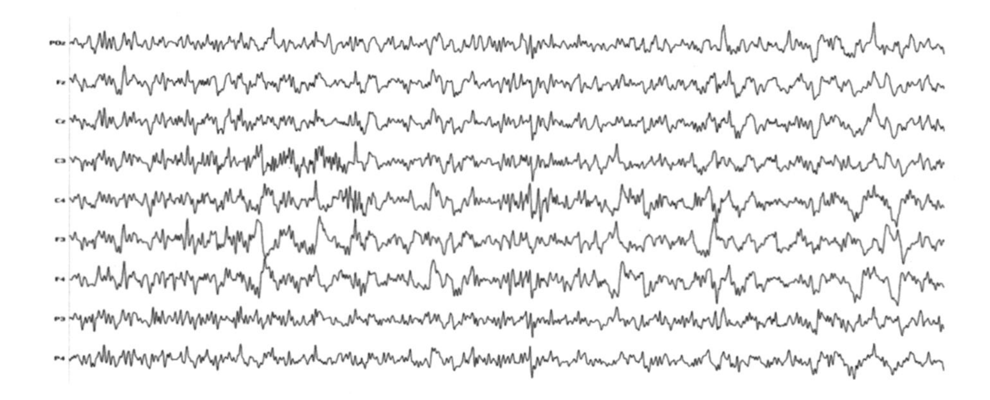

Fig. 3. The sample of human EEG with in resting state.

Human body movements or facial expressions affect the EEG record. Eyeball moves or limb moves cause artifacts directly on the EEG signal. The method of event-related potentials (ERP) returns brain response of a specific event. Figure 4 presents a signal reaction when the examined person clenched teeth three times - the signal amplitude grows.

The order of BCI stages is shown in Fig. 5. First, the procedure begins with signal acquisition when the device is connected to a computer. The processing signal consists of identifying features of the signal and then classifying the

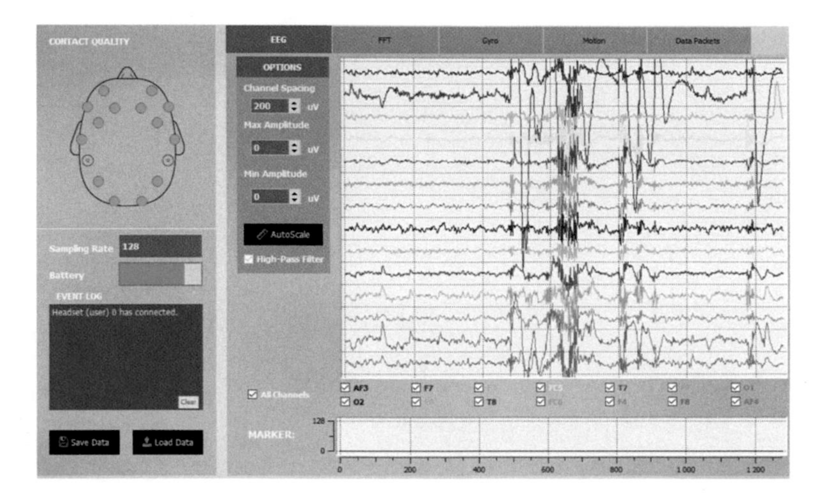

Fig. 4. The sample of human EEG with clenched teeth artifact - recorded in Emotiv Xavier TestBench.

selected features. With these measures the control signal is gained and then translated into commands. The received results determine the user's reaction which causes a feedback loop.

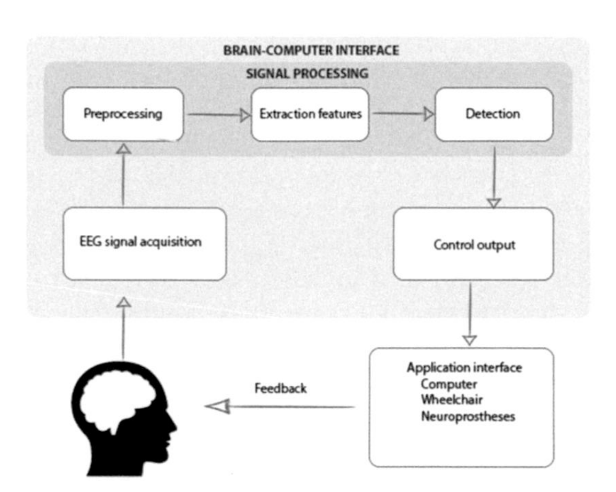

Fig. 5. The BCI system.

5 Picture Script

Picture script is used in a daily life. It is completely natural for their users because they have been accustomed to it since their childhood. Pictogram Ideogram Communication (PIC) is the simplest and most efficient method of communication with speech impaired persons or persons having difficulty to communicate with their family or friends. Pictogram symbols are monochromatic 10×10 cm pictures (Fig. 6). Every pictogram is signed on the top which helps understand the symbol by a person who does not know them and let persons who use pictograms remember letter writing. The set of pictograms consist of over 1000 signs divided into several categories. The imaging system is still developing and is being supplemented with new symbols adapted to specific regions or cultures of users [8].

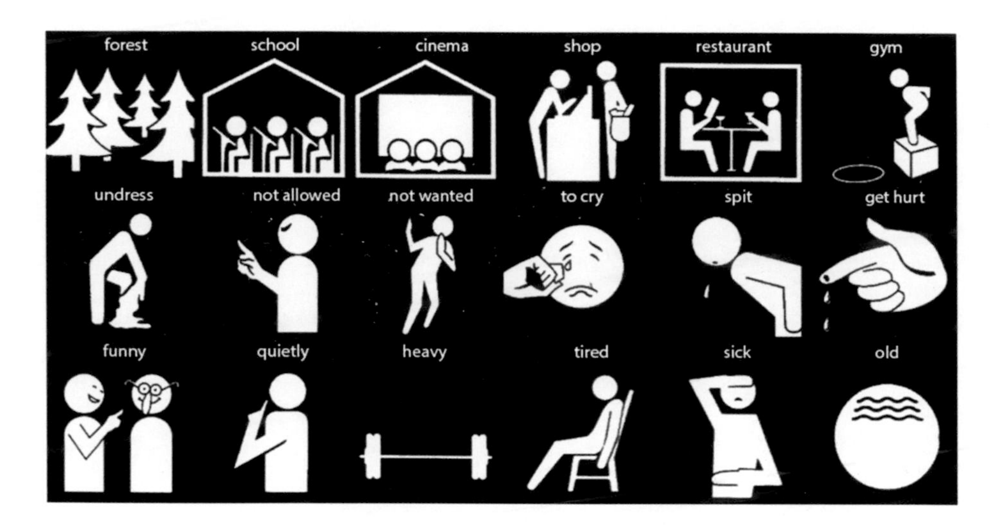

Fig. 6. Pictograms communication symbols.

Every pictogram represents a word or a term describing: an object, a person or an activity. The use of pictograms is very different - they can be applied as communication signs, activity plans, boards, daily plans, shopping lists, e-mails or even traditional letters. In everyday life picture script surrounds us in public zones as road signs, smartphone icons or symbols in public places. The alternative systems to verbal speech can perform two functions. Firstly, of course, making contact with other persons, conducting a conversation or expressing an opinion. Secondly, exploring the world. Thanks to communication systems humans can explore the environment and establish relationships with it.

6 System Concept

The aim of the research was to create a communication system for disabled people. The priority was to design minimalistic user interface based on a pictogram system. The application is specifically designed for a physically disabled person that is speech impaired. The speech therapist working with the disabled person uses pictogram boards for therapy. In practice, the person points the finger at the symbol. People who are not able to move an arm use a special sign to select a pictogram symbol - e.g. nod of the head or facial expression. The application functionality, interface design and controlling were consulted with a speech therapist who uses pictograms in her daily work. Particular symbols and categories implemented in this system are based on everyday needs of a disabled teenage girl with Dandy-Walker syndrome. This system requires the use of brain-computer interface EPOC+ Neuroheadset by Emotiv Inc. The device

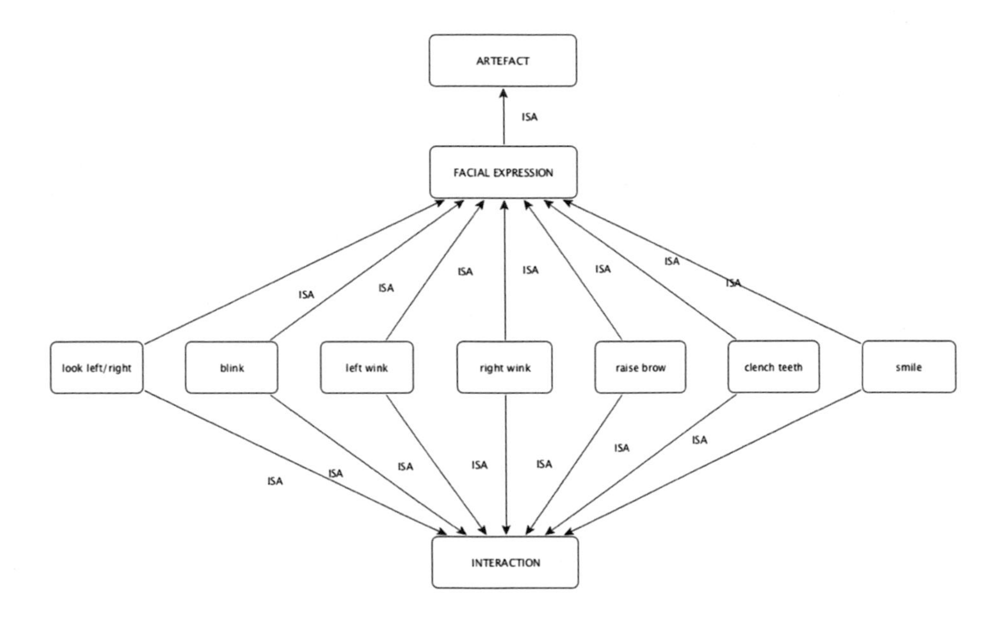

Fig. 7. BCI semantic network.

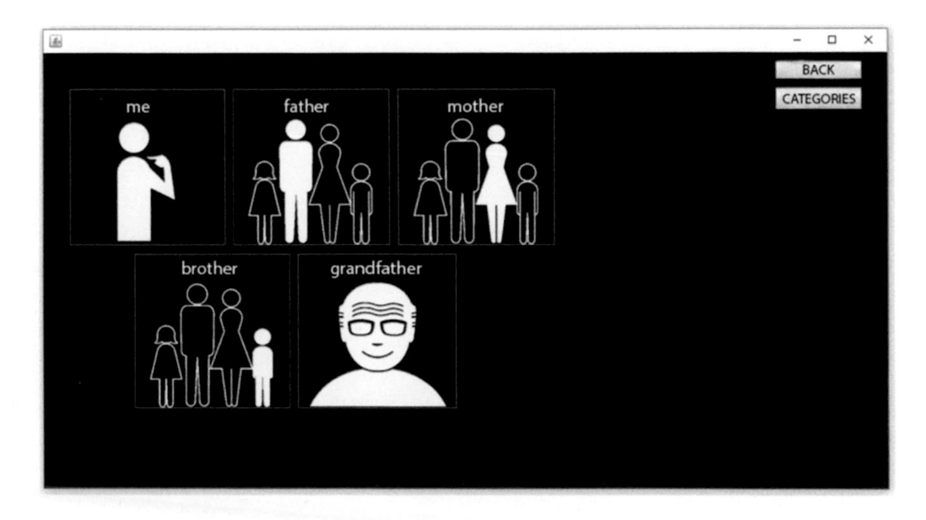

Fig. 8. Category panel: family.

placed on a head imitates a mouse cursor due to an embedded gyroscope. The facial expression (in this particular case clenching teeth) causes the left button click (Fig. 7).

The application is created in a Java programming language. The interface consists of over 30 categories which contain 4 or 5 symbols - e.g. in the doctor's office, weather, food or family (Fig. 8) [12]. This minimalistic form eliminates the loss of the user's focus and it also affects the low level of application difficulty.

7 Test Phase

The presented system complements a verbal communication. Depending on the type of disability, the user is able to operate the application by a touch screen, a keyboard, a computer mouse or a BCI device. In this particular case the author is using EPOC+ Neuroheadset device paired with a software developed by the Emotiv Inc. The first step was to activate gyroscope and set clench teeth as the left mouse button. Based on the mentioned earlier family category, the therapist can ask simple questions - e.g.: Who is waiting for you?, Do you have a sibling?, Who helps you get dressed today? and many others. To answer a question the user is moving her head which causes mouse cursor moves. When this one is placed on a selected symbol, the user has to confirm it by clenching her teeth. This kind of a nonverbal communication brings much more comfort into the disabled person's life.

8 Discussion

The system fully serves its purpose. The user interface is clear and user-friendly. It contains only the necessary functionalities. The people for whom it is dedicated require maximum reduction of all unnecessary functions or keys due to their limited needs. The permanent position of symbols allows the user to remember the arrangement relative to tabs or categories, which affects faster and more effective communication. Using Brain-Computer Interface in order to control the application (communicator) extends the scope of its use relative to different disabilities. It helps physically disabled persons and persons with speech impediment. It can also diversify a therapy in mental disability. The system can be a combination of all these features. This type of solution is the most efficient method of communication at this moment. It gives disabled people the possibility to become partially independent in their daily life. Brain-Computer Interfaces have huge potential, they are innovative and definitely they will expand on the market. It is also a great issue for further research.

References

1. Emotiv EPOC Specifications, document of Emotiv Inc.
2. Emotiv EPOC User Manual, document of Emotiv Inc.
3. Quick Start Guide, document of Emotiv Inc.
4. Paszkiel, S.: Interfejsy Mozg-Komputer. Neuroinformatyka. Of. Wyd. Pol. Op., Opole (2014)
5. Wee, K.H., Turicc, L., Sarpeshkar, R.: An articulatory speech-prosthesis system. In: Proceedings of the International Conference on Body Sensor Networks (BSN), pp. 133–138 (2010)
6. https://www.emotiv.com/consumer-insights-solutions/
7. Vecchiato, G., Toppi, J., Astolfi, L., Cincotti, F., De Vico Fallani, F., Maglione, A.G., Borghini, G., Cherubino, P., Mattia, D., Babiloni, F.: The added value of the electrical neuroimaging for the evaluation of marketing stimuli. Bull. Pol. Acad. Sci. Data Min. Bioeng. Tech. Sci. **60**(3), 419–426 (2012). https://doi.org/10.2478/v10175-012-0053-2
8. Błeszyński, J.: Alternatywne i wspomagające metody komunikacji, Of. Wyd. Impuls, Kraków (2006)
9. Rak, R., Kołodziej, M., Majkowski, A.: Brain-computer interface as measurement and control system. Pol. Acad. Sci. Metrol. Meas. Syst. **XIX**(3), 427–444 (2012)
10. Górska, M., Olszewski, M.: Interfejs mozg-komputer w zadaniu sterowania robotem mobilnym. PAR **19**, 15–24 (2015). https://doi.org/10.14313/PAR_217/15
11. Paszkiel, S.: Control based on brain-computer interface technology for video-gaming with virtual reality techniques. J. Autom. Mob. Robot. Intell. Syst. JAMRIS **10**, 3–7 (2016). https://doi.org/10.14313/JAMRIS_4-2016/26
12. Browarska, N.: Wykorzystanie urzadzenia Emotiv EPOC+ Neuroheadset do zautomatyzowanej komunikacji osob niepelnosprawnych z otoczeniem. Praca inzynierska, Politechnika Opolska, WEAiI, Opole (2017)

Methods of Acquisition, Archiving and Biomedical Data Analysis of Brain Functioning

Szczepan Paszkiel(✉) and Piotr Szpulak

Department of Biomedical Engineering, Faculty of Electrical Engineering, Automatic Control and Informatics, Opole University of Technology, Proszkowska 76, 45-271 Opole, Poland
s.paszkiel@po.opole.pl

Abstract. The following article sets out four acquisition methods of data obtained on the basis of brain signals: EEG, NIRS, fMRI as well as PET. Moreover, it provides the readout analysis of the signals occurring within the human brain and a possible manner of archiving and processing them. For an illustrative readout of the signals, a multi-channel encephalograph was applied. With the use of Emotiv Xavier TestBench application, time-varying EEG signals from individual electrodes were recorded in the .edf format which were subsequently subjected to Toolbox EEGLab for Matlab.

Keywords: EEG · NIRS · edf · Data analysis

1 Introduction

Currently, a dynamic increase in the demand of modern acquisition methods can be noted. It implicates development in the area of the biomedical engineering, which empowers the medical world with a growing number of equipment possibilities every year. Though, sole data collection is not the only possibility afforded by modern technology. The market of medical solutions offers more and more applications suitable for archiving and biomedical data analysis. In this view, the human brain seems to be the area of a particular interest, which provides basis of archiving huge collections of specific signals/images [1]. To carry out a quick and impressive patient's diagnosis, doctors more often exploit systems supporting their decisions in this sphere. Expanding data files gathered in data bases and warehouses require newer and challenging algorithms of exploration and inventing some behaviour benchmarks for the purpose of repetitive dependencies identification.

Szczepan Paszkiel, PhD. Eng., Assistant Professor; Piotr Szpulak, Msc. Eng.

W. P. Hunek and S. Paszkiel (Eds.): BCI 2018, AISC 720, pp. 158–171, 2018.
https://doi.org/10.1007/978-3-319-75025-5_15

2 Data Acquisition Methods

Clinical encephalography lays among the available acquisition methods of human brain data. It was introduced by Hans Berger, a German psychiatrist, in the 1930s. It belongs to the non-invasive methods involving both detection and registering electric brain functions with the use of electrodes distributed on the head surface and recording changes of electric potential on the skin surface originating from the activity of cerebral cortex neurons which are appropriately intensified prior to their recording-encephalography. The EEG records (Fig. 1) the following waves: Alpha, Beta, Theta, Delta, Gamma, Mu. The waves investigated during the tests are mainly beta ones with frequencies ranging from 13 to about 30 Hz and amplitude under 30 μV. They depict the cerebral cortex engagement into cognitive functions (hearing, sight, taste and touch). Small amplitude beta rhythms occur during attention concentration [2]. Additionally, they may be induced by various pathologies as well as chemicals such as benzodiazepines.

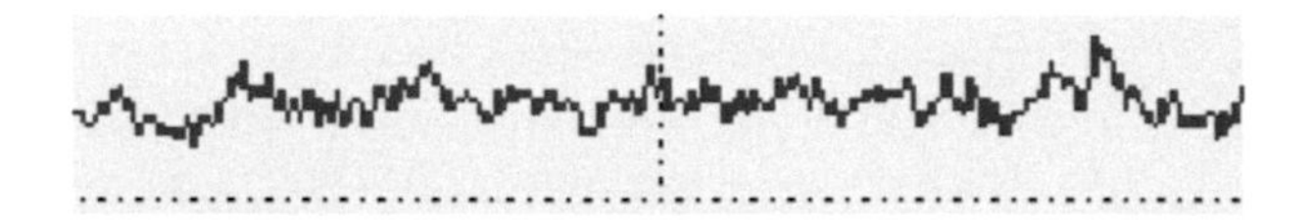

Fig. 1. An example of EEG signal recording in a function of time, horizontal axis time [ms], vertical axis amplitude [mV].

Near Infrared Spectroscopy (NIRS) serves as another technique of brain functioning visualization that involves laser rays to pass through a skull. Such lasers are very weak, thus they operate with a frequency of a light wave approaching an infrared value ranging from about 700 nm to 2500 nm, in the scope of which the scull becomes transparent. Blood containing oxygen absorbs light waves frequencies different from blood values, where oxygen has already been taken up. Therefore, observing the amount of light with various frequencies, it is possible for the scientists to monitor the blood flow. In case of activities mapping, such a concept is known as an optic tomography of Diffused Optical Technique (DOT). As far as registering of light dispersion is concerned due to the fluctuations generated during neurons stimulation, it is the Event-Related Optical Signal (EROS). It constitutes a brain scanning technique that applies infrared light through optical fibres for the evaluation of the optical features fluctuations within the active cerebral cortex areas. While such techniques as diffused optical imaging and NIRS measure the optical absorption of haemoglobin, thus basing on the blood flow, EROS makes use of the dispersion attributes of the neurons themselves, at the same time ensuring a significantly more direct measures of cellular activity. EROS can indicate brain activity in millimetres and milliseconds. Currently, lack of activity detection possibility for the depth beyond few centimetres displays its substantial limitation since it constraints a quick image processing within the cerebral cortex. Functional Magnetic Resonance

Imaging (fMRI) - constitutes the third specialized imaging mode with the use of magnetic resonance method which enables to measure the increase of blood flow as well as oxygenation of an active brain area. This method utilizes the fact that together with the nervous cells activity their oxygen demand rockets and carbon dioxide production is intensified. The increase in the activity of a specific area is measured with a Blood-Oxygenation-Level-Dependent response (BOLD) which defines the dependency of magnetic resonance signal intensity on the blood oxygenation leverage. The fMRI (Fig. 2) concept stems from the exploitation of MRI test, extending it to an observation based properties on the oxygenated and not oxygenated blood values. A subject undergoing the test is placed into a strong parallel-lines magnetic field. The coils built into the scanner transmit short electromagnetic impulses with a specified frequency towards the tested subject inducing the protons' spins inside the nuclei of hydrogen atoms which are mandatory components of water particles present in living organisms. In case of a static magnetic field with a value of 1,5 Tesla, the frequency is about 64 MHz. Due to the impulse operation, nuclei atoms are magnetized and become the electromagnetic field source themselves. After the impulse operation ceases, the electromagnetic radiation, generated by the spins returning to their de-energised state, is recorded by the coils functioning as receivers. Coming to the initial position, the protons' electromagnetic emission is decreasing with time with a frequency similar to the electromagnetic impulse sent towards them. The speed of wave disappearance relies on the characteristic properties of magnetic atoms inside individual tissues. Recording these waves, with the use of the so-called static magnetic field gradients, enables to recreate the image of the tested object interiors with a computer.

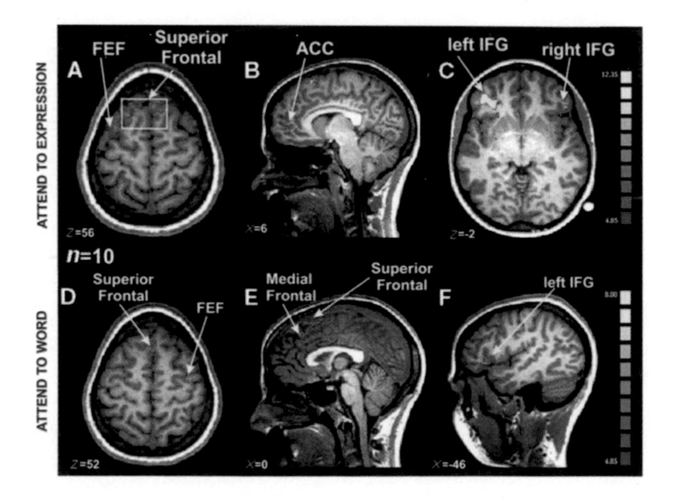

Fig. 2. fMRI test result depicting brain activities mapping [3], Anterior Cingulate Cortex (ACC), Left Inferior Frontal Gyrus (IFG).

Positron emission tomography (PET) is an imaging technique where instead of an outer X-ray or radioactive radiation source, as in computer tomography, the radiation generated during positrons annihilation is recorded (Fig. 3). A radioactive substance, undergoing beta plus decomposition, administered to a patient serves as the positrons' source. That substance contains radioactive isotopes of low half-life values allowing most of the radiation to occur during the test what in turn causes limitation of tissue damage due to radiation. It is also associated with a need to mobilise a cyclotron nearby that significantly increases costs. Nowadays, practically all the available positron emission tomography scanners are hybrid-type: PET-CT, PET/CT - a connection of PET with a multislice computer tomography, PET-MRI, PET/MRI - a connection of PET with magnetic resonance. Generated in the course of a radioactive decay and after passing a few millimetres way, the positrons collide with the electrons present in the body tissues, undergoing annihilation. As a result of this pair of electron–positron annihilation, two pieces of electromagnetic radiation quanta arise (photons) of 511 keV energy value each, moving in the opposite directions (at an angle of 180°). These photons are simultaneously recorded by two of many detectors set at various angles to a patient's body (most often in a ring form) which allows to determine the exact place of forming the positrons. The data, digitally registered onto a computer disk, enables to create the sectional images of a patient's body which are analogical to the images obtained with a magnetic resonance imaging method (3D image of a tested element is obtained).

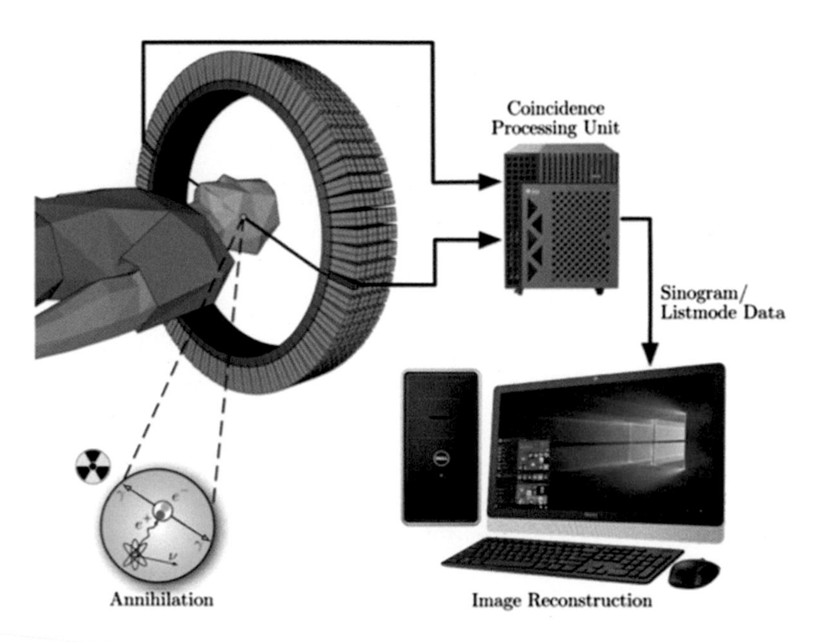

Fig. 3. Overall PET implementation pattern.

3 EEG Signal Artefacts and Data Signals Archiving

Disturbances distort the actual progress of brain impulses. They result from e.g.: eye movements, muscles pulsations, head movement etc. (biological artefacts). The impact of the results is directly proportional to the generated signal amplitude and inversely proportional to the distance between the sources and EEG electrodes. An insufficient contact of an electrode with a head surface belongs to significant sources of technical artefacts. In that case, minor head movements result in a decrease of an electrode adhesion to a skull causing a sudden, short rise of an electrode impedance. That outcerebral potential is registered by an electrode due to which it arises. Removing the disturbances may be facilitated, among others, either by means of relevant referential electrodes or by the received signal filtration (e.g. through low-pass or band-pass filters).

The data obtained in the course of the acquisition shall be properly stored with the aim of its further processing. The *.edf - European Data Format belongs to a very popular and standardized format. Such a format is also applied by the Emotiv Xavier TestBench application. The European data format is a standard file type intended to exchange and store medical time series. *.edf is commonly used for archiving, exchange and analysis of commercial appliances data in a format independent on the acquisition system. It permits the data to be downloaded and analyzed by an independent software. *.edf, introduced in 1992, stores multichannel data offering various sampling frequencies for each signal. *.edf format contains a heading and one or more data registers. The heading features some general information (patient identification, start date etc.) as well as technical specifications of every signal (calibration, sampling rate, filtration) designated as ASCII symbols. Data registers incorporate samples as 16-bites whole numbers.

4 Devices Used for EEG Data, Signal Archiving

The Emotiv EPOC+ NeuroHeadset device (Fig. 4) is produced by Emotiv Systems Inc. and, according to the manufacturer, it can read four mental conditions, 13 conscious conditions, mimics and head movements. The company operates since 2003 and specialises in BCI technique basing on EEG that is electroencephalography. The Emotiv EPOC+ NeuroHeadset exploits 14 biosensors which may test brain waves activity after the saturation with saline solution.

The appliance is compatible enough to operate with Windows, Linux, MAC OSX, Android and iOS systems. It has an in-built Intel Premium 4 processor, 2 GB RAM and USB port 2.0. The battery working time is supposed to be 12 h. A number of electrodes located in appropriate spots over the head surface register the electrical brain activity. The sensor areas are determined according to professor Jasper's scheme from 1958 - 10–20 system. The given figures correspond to 10 or 20% distance values of three segments designated on the head surface due to standard orientation points (sagittal line, coronary line and a transverse line). Together with the implementation of the Emotiv EPOC ControPanel application it was possible to configure a connection between the Emotiv EPOC+ NeuroHeadset and a computer. The result was presented in Fig. 5. Green colour depicts the correct correlation of an electrode with the head skin.

Fig. 4. The Emotiv EPOC+ NeuroHeadset appliance.

Followingly, with the use of Emotiv Xavier TestBench application it was possible to observe the behaviour of signals obtained from the biosensors in the testee's relevant response. Neurons' behaviour of a testee at rest was presented in Fig. 6. It can be noted that the signals oscillate around a relatively low amplitude at standstill. Correctly connected biosensors are visible in this picture on the left. Two of them, marked with grey-green (placed behind an ear) are reference detectors which serve as measurement benchmarks for further sensors.

Responses to slight movements of eye lids are depicted in Fig. 7. Arrows indicate the spots where neuron response for closing a left or right eye is visible. Disturbances occurring due to the testee's reaction can also be noted, when the right eye simultaneously made a minor movement while the left eye was being closed.

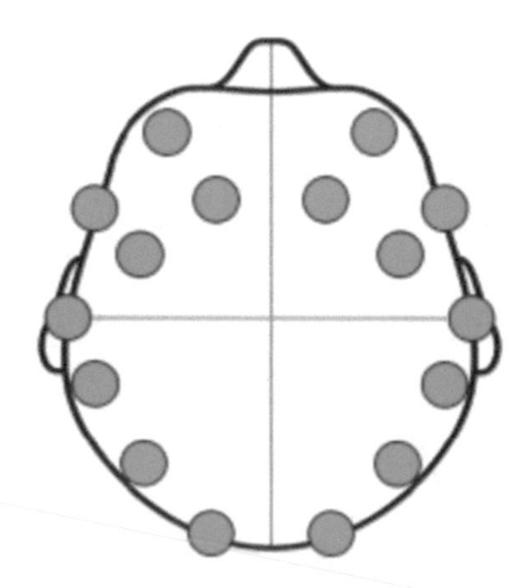

Fig. 5. Configuration of the Emotiv EPOC+ NeuroHeadset.

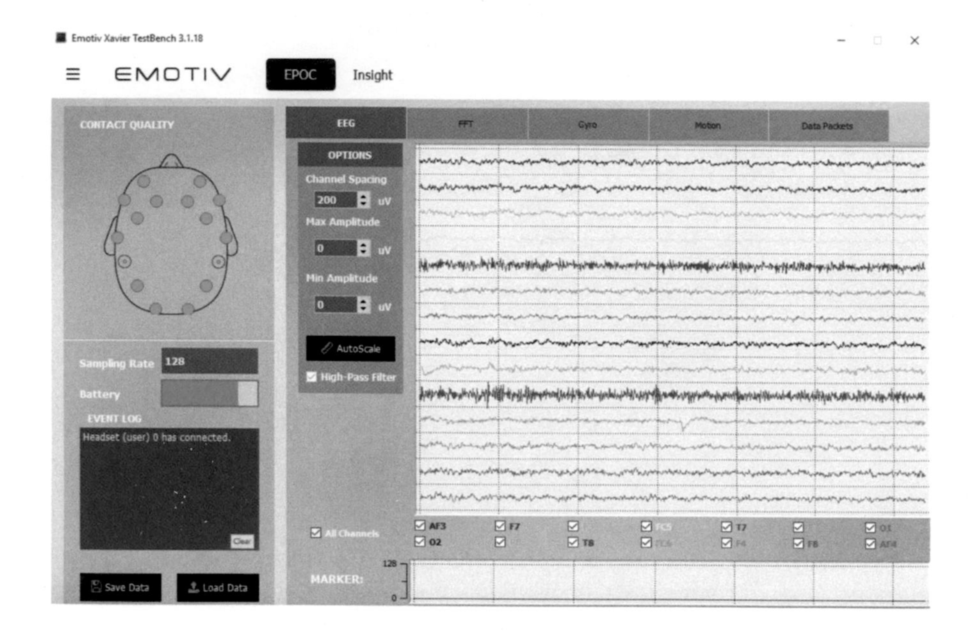

Fig. 6. Standstill condition.

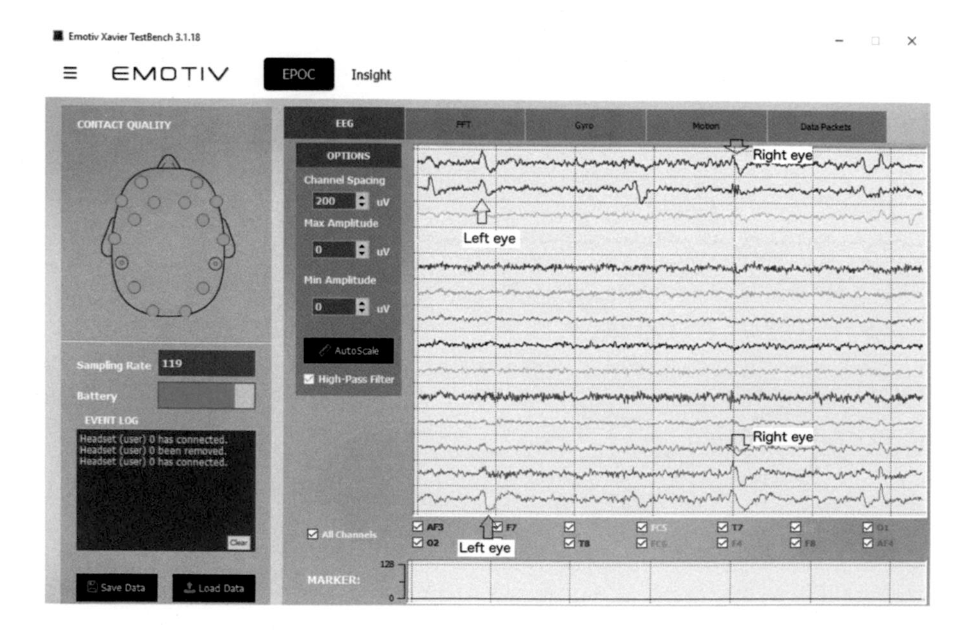

Fig. 7. Minor eye movements.

5 Data Analysis in EEGLAB for Matlab

The data acquired in the Emotiv Xavier TestBench environment may be used for further processing. The data format (*.edf) is compatible with Toolbox EEGLAB [4]. It is an interactive tool within the Matlab environment for processing continuous data connected with EEG, MEG and other electrophysiological events covering Independent Components Analysis (ICA), time analysis, frequency and artefacts removal [5]. EEGLAB operates under Linux, Unix, Windows and Mac OS X systems. Toolbox EEGLAB is not available in the default version of Matlab, therefore it must be installed. Figure 8 presents a EEGLAB configuration window, where data processing runs. Figure 8 also delineates an access path for reading the obtained data.

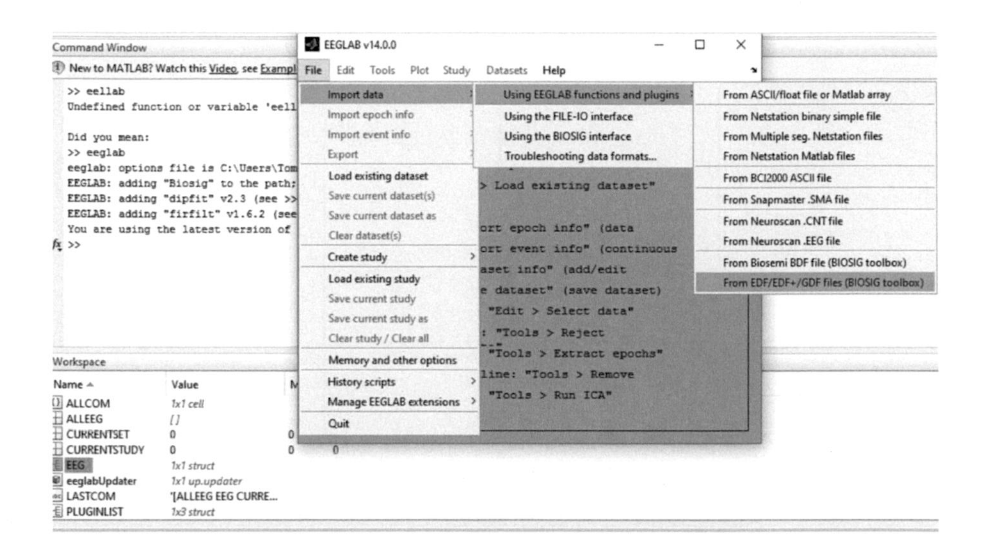

Fig. 8. Toolbox EEGLAB window for Matlab.

It can be noted that a structure occurred within the working space where all the parameters are incorporated. Raw data itself is stored in the matrix form (Fig. 9), for which the data from subsequent sensors is provided in the following verses and the data from subsequent time spans - in columns.

It is easy to observe that the data in Fig. 10 is complex for identification and analysis. There are multiple disturbances (artefacts). Thus, the EEGLAB environment enables to process the data in order to gain e.g. only the specified signal frequencies. It is widely known that the beta waves frequency ranges from 13 to 30 Hz, therefore a band-pass filter has been applied for the aim of receiving only this signal on the individual electrodes. The configuration of this filter was presented in Fig. 11.

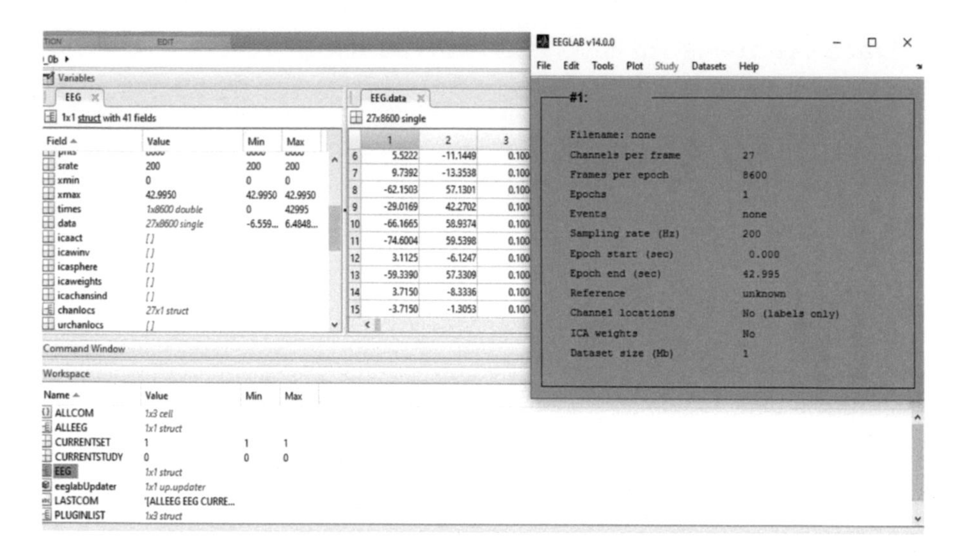

Fig. 9. Input data - Matlab.

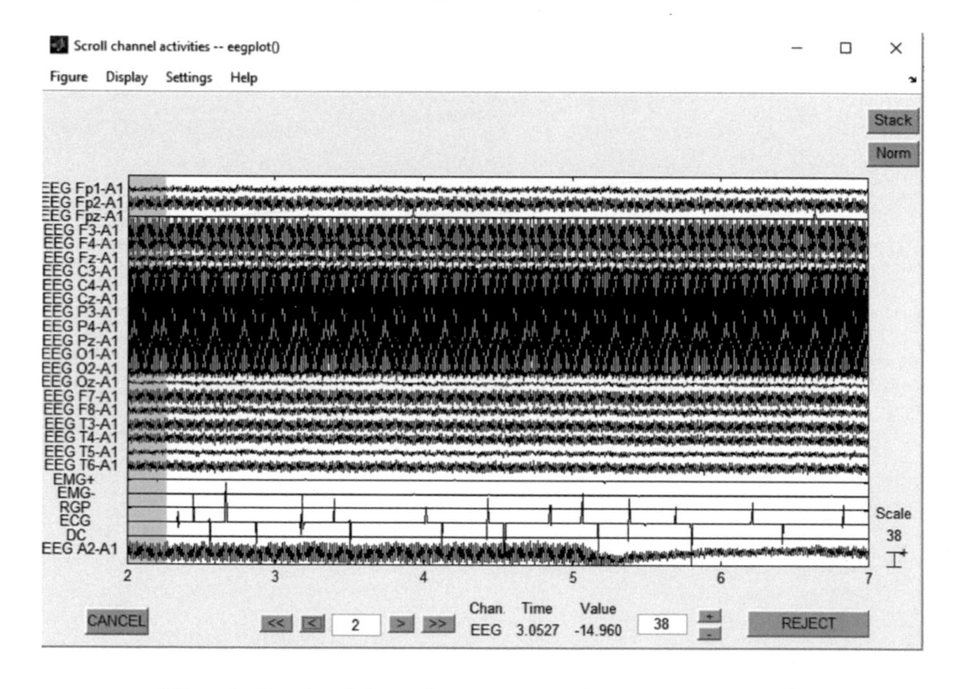

Fig. 10. Received data characteristics (without processing).

Deployment of this filter causes another data object to occur with the information generated as a result of passing a primary signal through the filter (Fig. 9).

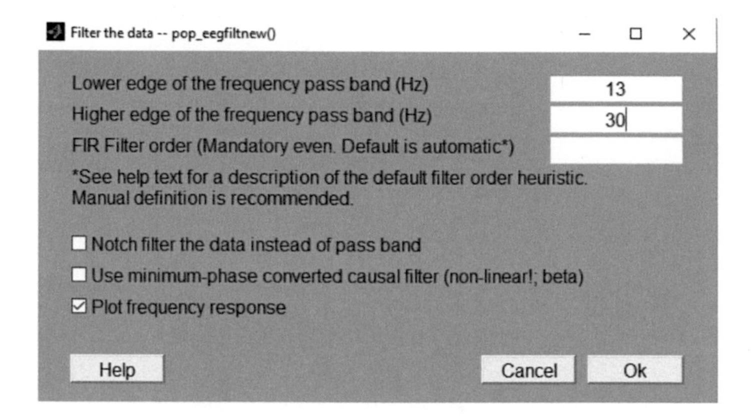

Fig. 11. Band-pass filter configuration for f: 13–30 Hz.

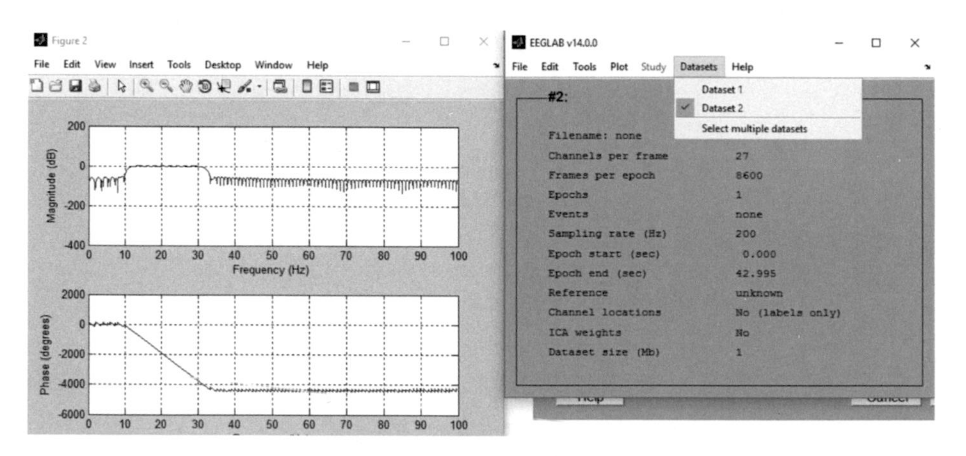

Fig. 12. Band-pass filter for f: 13–30 Hz.

Figure 12 also represents Bode's plot characteristics of the applied filter. It pinpoints that for the 13–30 Hz range, both the module and the signal phase undergo fluctuations.

Figure 13 proffers the received signal transformed by the filter. As a result of comparing Figs. 10, 11, 12 and 13, it is highlighted that a band-pass filter serves as an excellent tool for filtration of artefacts and disturbances influencing the received signal. The data generated in that way may be helpful to conduct various processes. Figure 14 uncovers further exemplary information. It may be noted that in the working space a new structure appeared where all parameters are present. Raw data itself is stored in the matrix form, for which the data from subsequent sensors is provided in the following verses and the data from subsequent time spans - in columns.

Figure 15 depicts heavily distorted data. It involves multiple disturbances (artefacts). Thus, The EEGLAB environment enables to process this data to

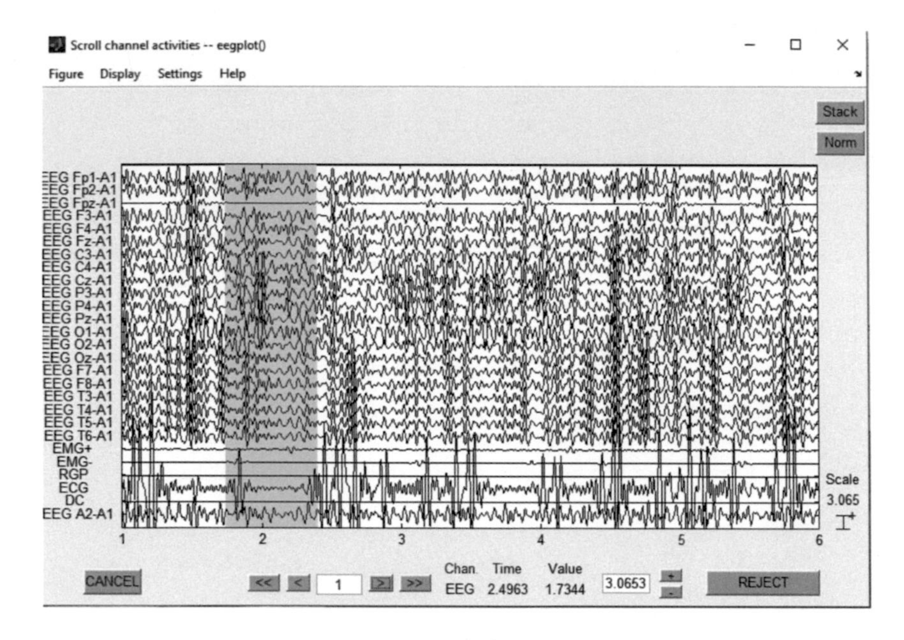

Fig. 13. Characteristics of the received data (after filter verification).

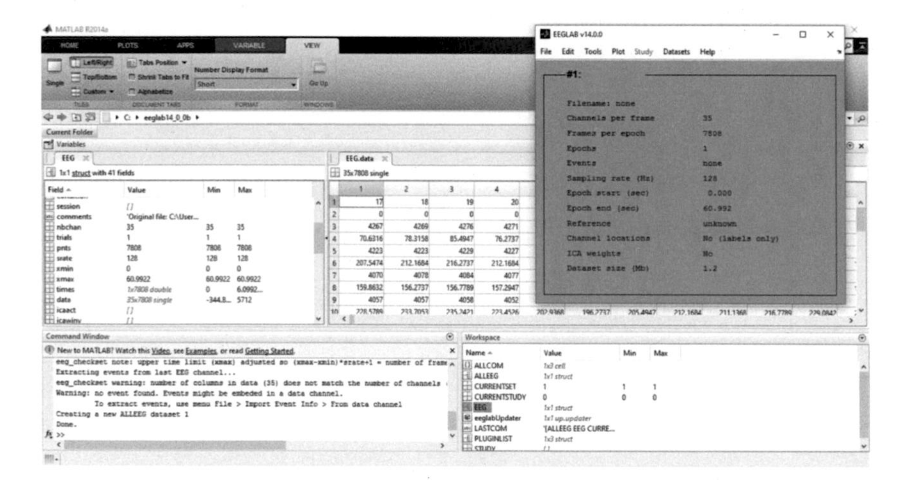

Fig. 14. Subsequent input data in Matlab environment.

acquire e.g. only the specified signal frequencies. As it has already been noted, beta waves range between 13 and 30 Hz, and therefore a band-pass filter has been applied to acquire this signal only on the individual electrodes.

The application of this filter prompts to generate another object with the data that were recorded due to a primary signal conversion (Fig. 15) through the filter.

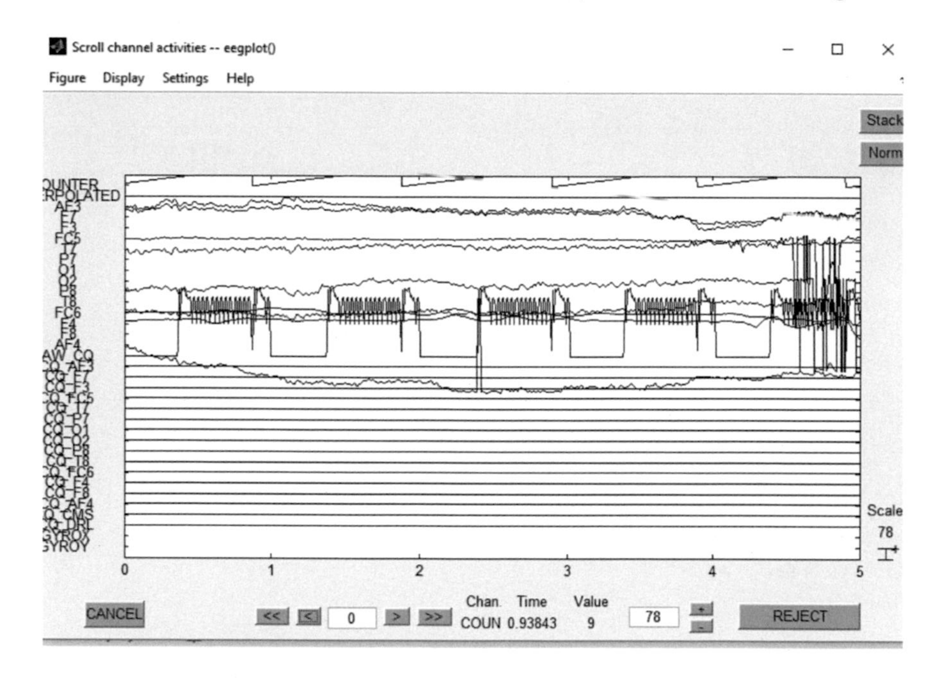

Fig. 15. Characteristics of the received data (without processing).

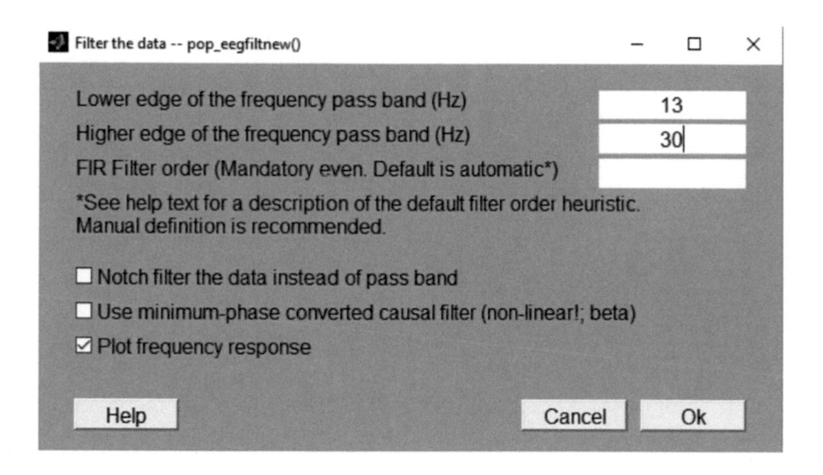

Fig. 16. Band-pass filter configuration for f: 13–30 Hz

Figure 17 also provides Bode's plot characteristics of the applied filter. Figure 18 presents the received signal after passing the filter. As a result of comparison of Figs. 15, 16, 17 and 18, it is possible to delineate an excellent feature of the band-pass tool to filter artefacts. In both cases, the band-pass filter of 13–30 Hz frequency values enabled to receive relatively "clear" signals which might be further used for processing or for application in a control process [6].

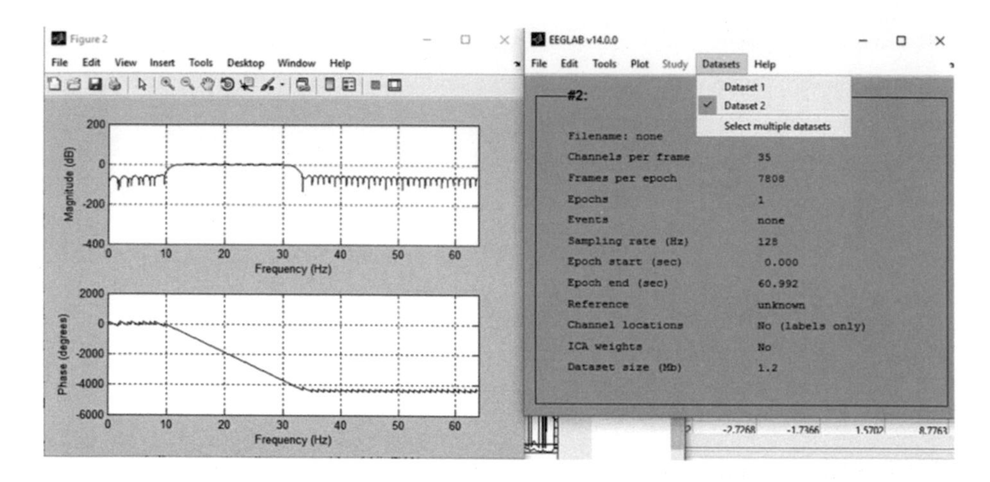

Fig. 17. Band-pass filter for f: 13–30 Hz.

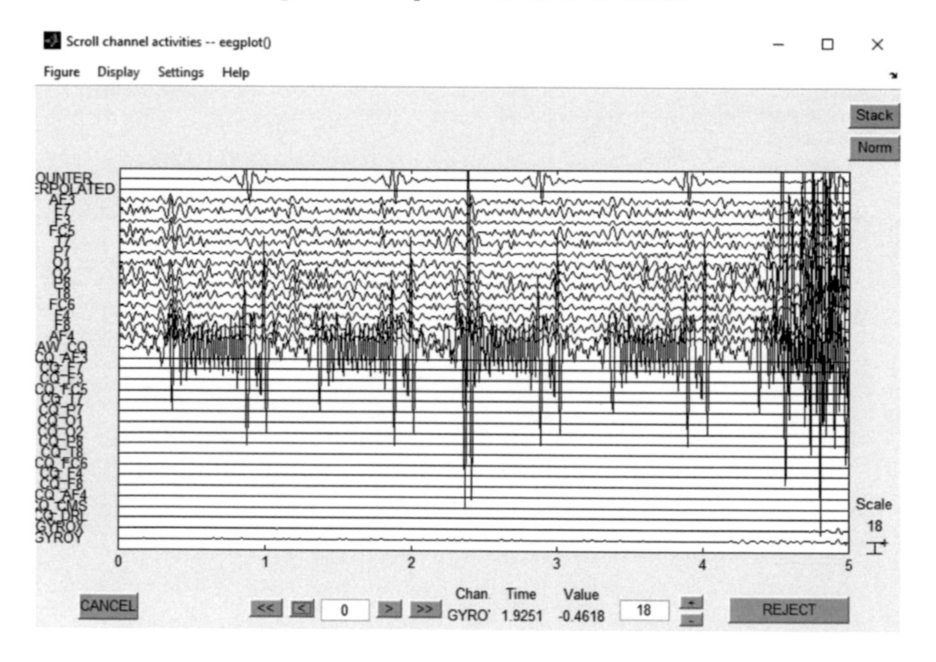

Fig. 18. Characteristics of the received data (after filter processing).

6 Discussion

The devices basing on the Brain-Computer Interface (BCI) technology allow for exchange of the brain signal into the digital one, which is transmitted to various kinds of instruments controlling computer applications and electronic equipment (e.g. wheelchair remote) without any muscle intervention [7]. An important issue of that aspect is connected with the fact that the *.edf format is widely applied through standardisation (it is not the only format of data archiving, however it belongs to a very commonly used one), what results in rapid and increased development of devices produced by different companies as well as their relatively better synergies and compatibilities. Due to the theory exploitation, including knowledge of individual waves frequencies, among others, and appropriate filters reducing artefacts, a greater volume of human brain activity conditions seems to be read with progressive precision. Having achieved the exact data concerning brain functioning of e.g. a person suffering from epilepsy, through relevant stimulation of specific neurons, it will theoretically be possible to entail limitations of some negative consequences stemming from that illness or even to reduce the prevalence of the illness itself.

References

1. Mathewson, K.E., Lleras, A., Beck, D.M., Fabiani, M., Ro, T., Gratton, G.: Pulsed out of awareness: EEG alpha oscillations represent a pulsed-inhibition of ongoing cortical processing. Front. Psychol., February 2011. https://doi.org/10.3389/fpsyg.2011.00099
2. Ghaemi, A., Rashedi, E., Pourrahimi, A.M., Kamandar, M., Rahdari, F.: Automatic channel selection in EEG signals for classification of left or right hand movement in BCI using improved binary gravitation search algorithm. Biomed. Sig. Process. Control **33**, 109–118 (2017). https://doi.org/10.1016/j.bspc.2016.11.018
3. Ovaysikia, S., Tahir, K.A., Chan, J.L., DeSouza, J.F.X.: Word wins over face: emotional Stroop effect activates the frontal cortical network. Front. Hum. Neurosci., January 2011. https://doi.org/10.3389/fnhum.2010.00234
4. Delorme, A., Makeig, S.: EEGLAB: an open source toolbox for analysis of single-trial EEG dynamics including independent component analysis. J. Neurosci. Methods **134**(1), 9–21 (2004). https://doi.org/10.1016/j.jneumeth.2003.10.009
5. Ghaemia, A., Rashedia, E., Mohammad, P.A., Kamandara, M., Rahdaric, F.: Automatic channel selection in EEG signals for classification of left or right hand movement in Brain Computer Interfaces using improved binary gravitation search algorithm. Biomed. Sig. Process. Control **33**, 109–118 (2017). https://doi.org/10.1016/j.bspc.2016.11.018
6. Paszkiel, S., Hunek, W., Shylenko, A.: Project and simulation of a portable proprietary device for measuring bioelectrical signals from the brain for verification states of consciousness with visualization on LEDs, Recent research in automation, robotics and measuring techniques. In: Szewczyk, R., Zielinski, C., Kaliczynska, M. (eds.) Challenges in Automation, Robotics and Measurement Techniques. Advances in Intelligent Systems and Computing, vol. 440, pp. 25–36. Springer, Cham (2016). https://doi.org/10.1007/978-3-319-29357-8
7. Wei-Yen, H.: Brain-computer interface connected to telemedicine and telecommunication in virtual reality applications. Telematics Inform. **34**(4), 224–238 (2017). https://doi.org/10.1016/j.tele.2016.01.003

Making Eye Contact with a Robot - Exploring User Experience in Interacting with Pepper

Michal Podpora(✉) and Agnieszka Rozanska

Faculty of Electrical Engineering, Automatic Control and Informatics,
Institute of Computer Science, Opole University of Technology,
ul. Proszkowska 76, 45-758 Opole, Poland
m.podpora@po.opole.pl, a.rozanska@gmail.com,
http://we.po.opole.pl/

Abstract. In this paper authors describe a new approach to analysis of user experience in interacting with a humanoid robot Pepper. The designers and engineers of humanoid robots struggle to make the robots appear friendly and welcoming – not only to surpass the so-called Uncanny Valley, but also to make the human-machine interaction consistent and comfortable. The paper includes description of an experiment involving humanoid robot Pepper and human volunteers in a form of a short verbal interaction. The result of the experiment consists not only of a questionnaire, but also other parameters measured and calculated by a robot, inter alia the time of the eye contact. The authors evaluate the correlation between the time of eye contact and the overall opinion of volunteers.

Keywords: Pepper robot · Robot · Humanoid · Uncanny valley
HMI · Eye contact · User experience

1 Introduction

The genesis of robots dates back to the ancient China [1]. Robots, especially those that resembled animals and humans, were built to fulfill a specific movement, behavior or function. Some of them were built to write, others to play music. Robots that resemble humans are nowadays called humanoids, whilst robots that appear nearly impossible to distinguish from a human, are called androids. In this paper, authors will focus on a specific humanoid robot, named Pepper.

Pepper is a humanoid robot introduced in 2014 by Aldebaran Robotics (now SoftBank Robotics) [2]. It is designed to make contact with a human as natural as possible [3]. Pepper's purpose is said to enhance people's lives and "to make people happy" [4]. It is able to localize a person talking to it, distinguish multiple faces, determine eye contact or even recognize and react to basic emotions of the person it is talking with. Voice-based human localization is possible due to four directional microphones installed inside Pepper's head.

W. P. Hunek and S. Paszkiel (Eds.): BCI 2018, AISC 720, pp. 172–183, 2018.
https://doi.org/10.1007/978-3-319-75025-5_16

Pepper is able to recognize someone's emotion not only by voice, but also by parameterizing facial expressions of interlocutors by using machine vision. Pepper has two RGB HD cameras: one in the "mouth", and the other one on its forehead – seen as a black dot. The robot is also equipped with a three-dimensional Asus XTION depth sensor [5], installed inside one of its eyes. Also other, more physical, interactions with Pepper can be made – its design includes three touch sensor regions on Pepper's head and two more on its hands. A touch-capable display (tablet) is located on its chest, and it can also be easily used as one of the means of communication in the overall Human-Machine Interaction (HMI) system. The tablet supports showing images and other media content on its 10.1-inch display, as well as the ability for human to interact (e.g. by choosing some options, pressing virtual buttons).

The robot itself can actively interact with human not only by generating voice (using its built-in text-to-speech engine) or by showing content and options on the tablet, but also by using gestures. Pepper's body design includes twenty motors and three omnidirectional wheels that enable wide range of movement for the robot. It has a built-in anti-collision system to detect people and obstacles in its way to reduce the risk of collisions. Pepper is able to maintain its balance using its inertial unit, including 3D gyroscope and 3D accelerometer.

2 Programming Pepper

Pepper can acquire new functionalities by implementing packages for the NAOqi Operating System preinstalled in the robot. Programming Pepper can be performed in two ways: by using graphical interface (see Fig. 1) and/or in Python programming language (Fig. 2).

Using the Choregraphe environment, a simple background module application was implemented in order to evaluate time interval of eye contact as well as to "remember" the identity of a particular interlocutor. The application is presented briefly in following chapters.

The Choregraphe environment, presented in Fig. 1, offers the possibility to design, implement and run custom packages/applications (behaviors, interactions) using a simulator or a physical robot. The center part of the Choregraphe's interface is used for graphical programming of the robot? by putting specific action blocks and connecting their inputs and outputs. On the bottom left there are ready-to-use preprogrammed blocks for specific movements or sensors of the robot. Right side of the interface includes simulator view of the current robot pose/movement and a memory window for viewing values of shared variables. Designing a simple package/application for a robot is fairly easy, but for advanced (non-trivial) functionalities, behaviors and algorithms, developer has to use Python programming language. This can be done in two ways: either by modifying the underlying Python code of any of the preprogrammed blocks (double-clicking a block brings up a simple script editor – Fig. 2) or by using any external IDE (integrated development environment).

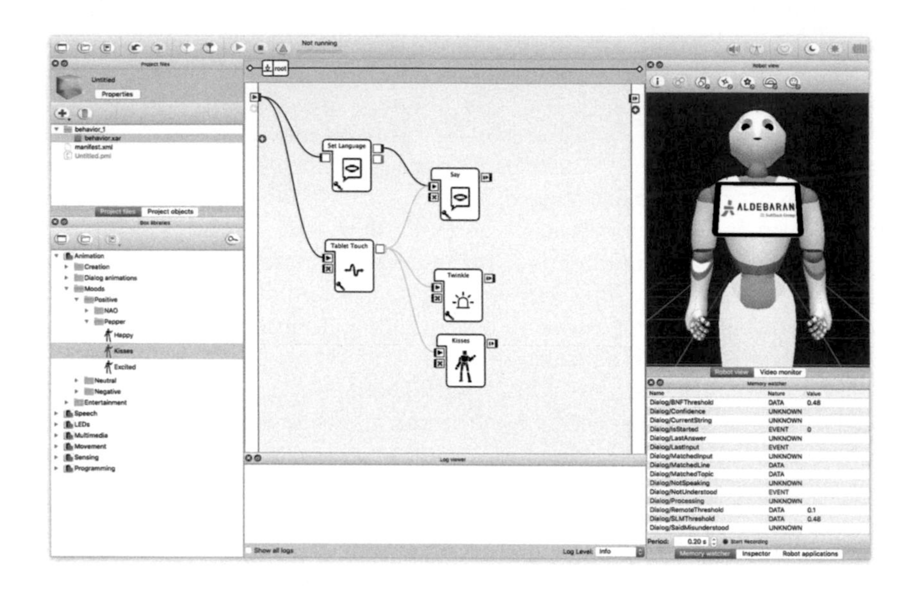

Fig. 1. The Choregraphe – GUI for graphical programming of a robot

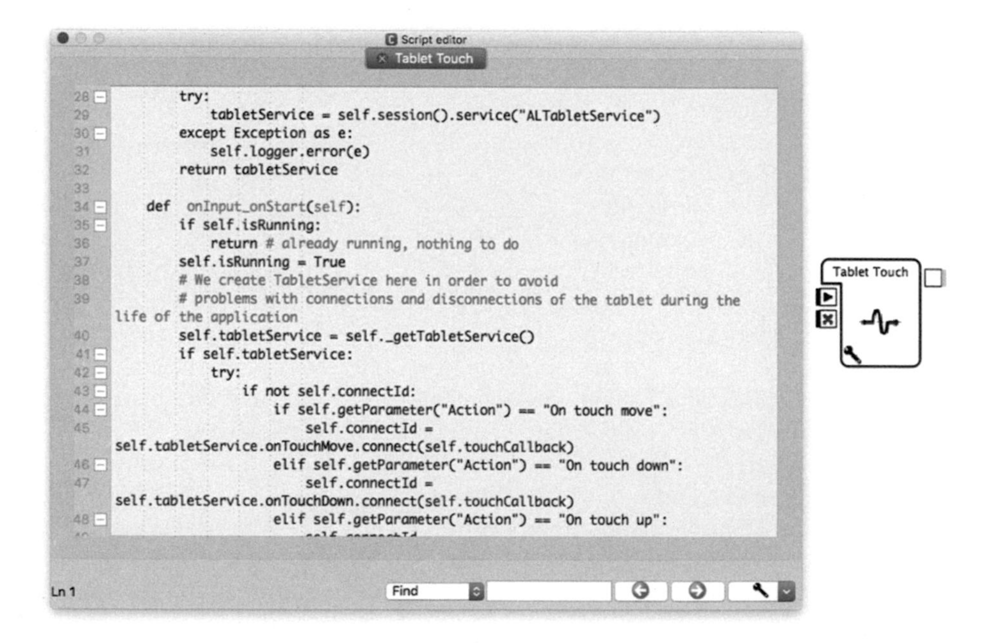

Fig. 2. Programming a robot in Python

3 Parameterizing User Experience – Testing Method

Pepper's NAOqi Application Programming Interface (API) includes a wide variety of preprogrammed modules and methods so that programming Pepper's behavior would be easier and the interaction would include more complex actions and responses. One of the modules useful for efficient and seamless user interaction and experience is the PeoplePerception module, which enables the possibility to detect and recognize faces. The PeoplePerception module includes the ALGazeAnalysis API, which offers special events that are being triggered whenever a human (present in the robot's field of vision) starts or stops looking at the robot. These events (ALGazeAnalysis/PersonStartsLookingAtRobot() and ALGazeAnalysis/ PersonStopsLookingAtRobot()) were used. Figure 3 presents the diagram view of a simple Choregraphe application implemented for the research.

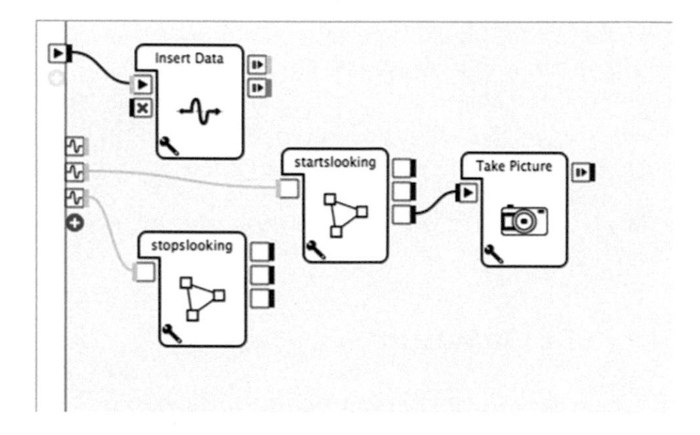

Fig. 3. The application for detection and measurement of eye contact duration (one "Take Picture" block and three custom Python blocks)

The application was configured to be executed automatically upon booting the robot, and then to continue to run in the background, regardless of other applications and behaviors. The very first block of the application, labeled "Insert Data", initializes shared memory resources and variables, using the ALMemory module. It is being triggered only once, when the robot's operating system triggers the "onStart" Input, indicating the successfully finished bootup sequence of the robot. Below the "input" icon (with black "play" triangle) there are three icons with a wave drawing. These represent events. In first attempts, authors had tried to use ALBasicAwareness/HumanTracked event, but it turned out that the robot was able to detect a human far before it was possible to distinguish who the person was. For this reason this event is not being used in the current version. The second event, visible in the Fig. 3, is the ALGazeAnalysis/ PersonStartsLookingAtRobot event [6]. It triggers the "startslooking" block, containing code for saving face ID number (called id_humana) in a shared memory variable, as well as the exact time when that person engaged the eye contact with the robot. After that, Pepper takes a picture containing the face of the person. Photos are taken only for the purpose of linking a particular person and their ID number. The photo is then saved as *.jpg file with the person's ID used as a file name. The resolution of the photo is 640×480 pixels. The third event, ALGazeAnalysis/PersonStopsLookingAtRobot, triggers the "stopslooking" block, which also checks the system time and calculates milliseconds between engaging the eye contact and its loss, and the time in milliseconds is saved into a text file with corresponding ID.

After the interaction with Pepper, the person is being interviewed or completes a survey. The survey includes several questions, presented below:

1. What is your age and gender?
2. How often do you use internet?
3. What is your first impression of communication with (this) robot?
4. Do you think (this) robot would be useful as a companion?
5. What makes (this) robot trustworthy?
6. What makes you comfortable/uncomfortable in interaction with (this) robot?
7. How do you feel about (the) robot looking at you? Does the eye contact make you feel uncomfortable?

The results of the completed tests and the survey are presented and commented in the next chapter.

4 Procedure and Quantitative Study

The application was active for 23 days of Pepper's "everyday life". During this period, Pepper interacted 763 times (over 95% of interactions took place in 10 of those 23 days). The authors did not interfere with the interaction between Pepper and any of the interlocutors – neither the content of nor quality of interaction, not to alter the user experience. Nearly half of the photos were blurry and it was not easy to recognize who (or which) was the interlocutor. However, most of the photos were sharp and authors were able to recognize the person, talk with them and complete the survey.

All of the people that interacted with Pepper and completed the survey are familiar to technology – they use internet daily. Their age ranges from 24 to 56 years (with median value at 29).

The graph below (Fig. 4) presents the eye contact duration time data collected by Pepper, only for those interlocutors who agreed for the interview/survey, sorted ascending. The graph is followed by visualizations of the survey answers regarding the overall user experience of the interaction and robot (Figs. 6, 7, 8, 9 and 10).

Figure 4 clearly shows, that there are more of the shorter duration recorded than longer, however it might be surprising that the shortest one was 1630 ms and that there is no shorter one. It seems to be intuitive that the graph should

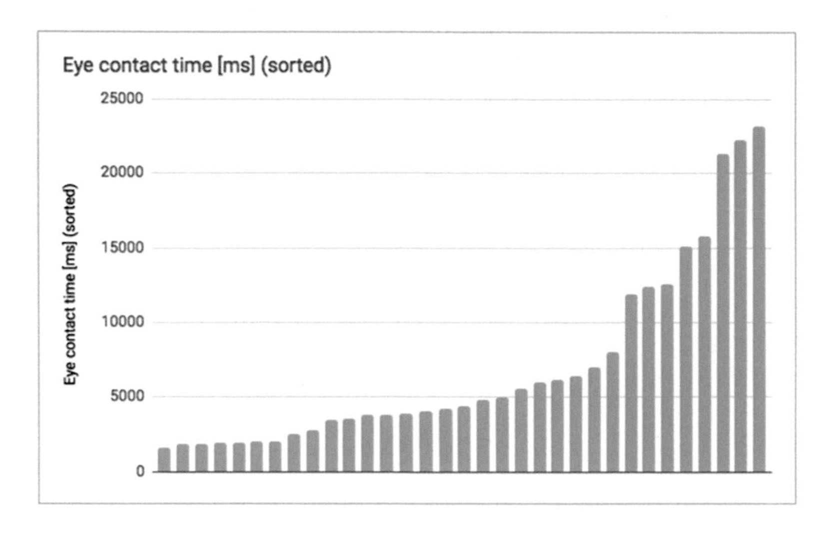

Fig. 4. Eye contact duration time data

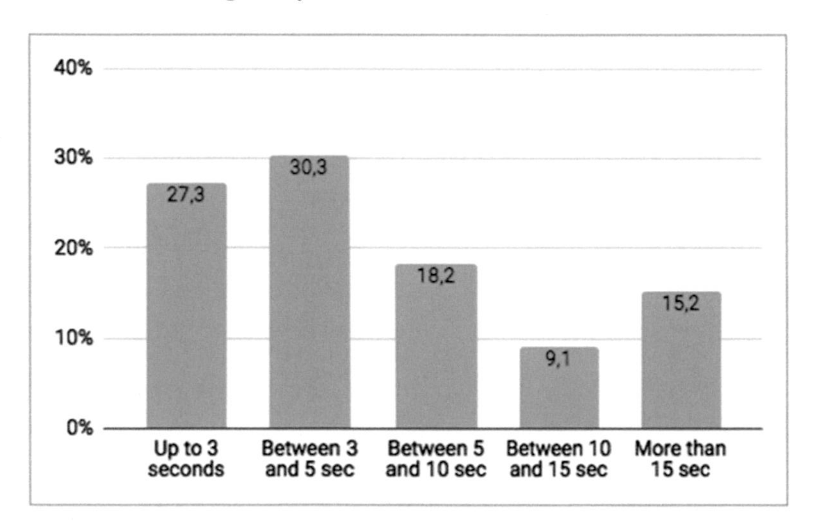

Fig. 5. Eye contact duration time data, grouped in ranges

start with one-frame (1/30 s) "saccades", than longer "glances", and finally eye contact measured in more than 1000 ms. The reason for not recording anything shorter than 1630 ms may be limitations of the proposed methodology – indeed there were a lot of shorter duration times, but there was no readable image corresponding to those recordings. The authors' interpretation for this is that the ALGazeAnalysis library may require some time to detect and report an event of a person actually looking at the robot.

Graph below (Fig. 5) shows eye contact duration grouped in ranges – for better interpretation of the graph shown in Fig. 4.

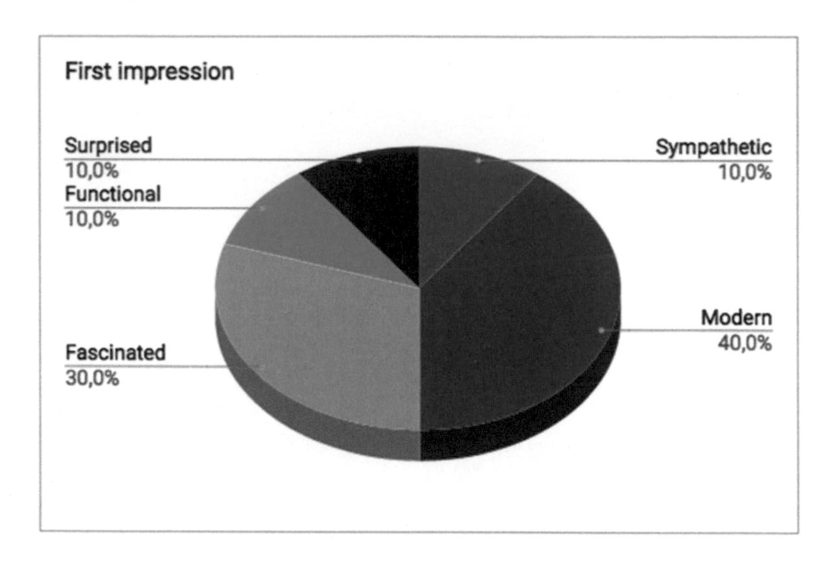

Fig. 6. Answers to the "first impression" question in questionnaire

Keeping in mind that the first bar ("under 3 s") of the Fig. 5 should be higher, the graph shows similar tendency – shorter duration. However, the last bar seems to be an exception – it is longer than the preceding one. This can also be seen in the Fig. 4 – there are some results which seem to stand out, which seem to form a group of instances, the are surely not incidental. The authors have investigated this issue and the explanation has been found – the robot's active "foreground" application for interacting with human involved a possibility of telling jokes, and the longest eye contact duration (Fig. 4) were similar to the duration of the jokes. It is also an interesting finding, that people (at least some of them) can establish longer eye contact with the humanoid and they do not look away for any reason.

The figures below present the information collected during the interview (and/or using the questionnaire). The first graph (Fig. 6) presents answers for question regarding interlocutor's first impression of the robot as well as its appearance/behavior (before first interaction). The answers were quantized and grouped to be represented as single adjectives.

As seen in the Fig. 6, most of the people said, that either they were fascinated about the robot or they consider it to be a modern technology. It can only prove that people nowadays are not afraid of robots and that the society becomes ready for a true coexistence of robots in our lives. The terms "modern" and "fascinated" are undoubtedly positive expressions, which gives a hint on how to improve future robots to be even more advanced (modern and fascinating). Other answers, such as "surprised", "sympathetic" and "functional", are also positive, which also proves the point.

The next graph – shown in Fig. 7 – presents answers to a question if the robot would be useful as a companion. This question was asked to see, if today's

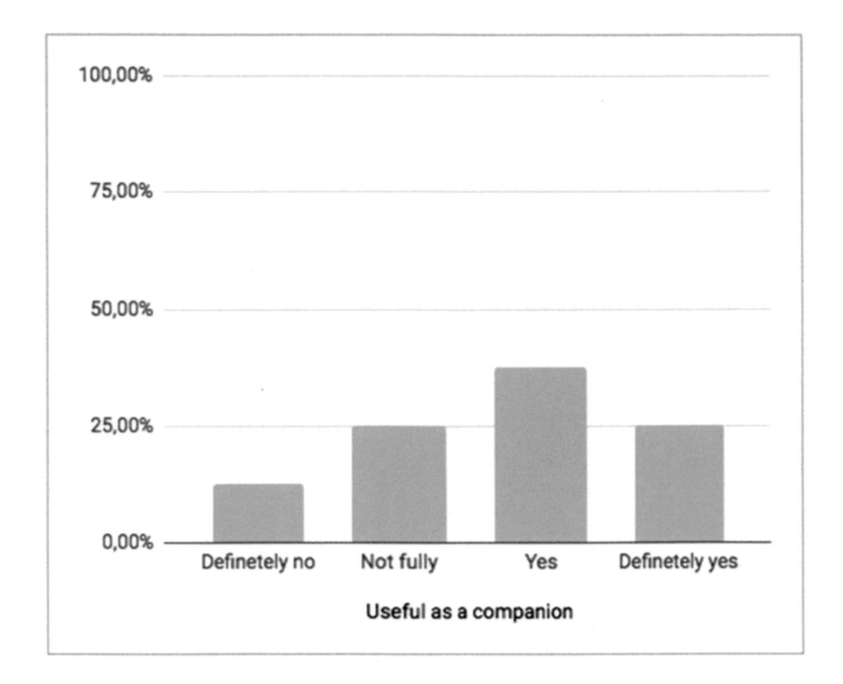

Fig. 7. Answers for a question if robot will be useful as a companion

society would find a humanoid robot trustworthy, and thus it could be treated as a companion.

In Fig. 7, first bar is significantly lower than the rest of the answers. Most of the answers were yes and definitely yes, which can be seen in the right side of the graph. The results can lead to a conclusion that most of the people have a positive opinion about robot being a companion.

Figure 8 presents answers for question "what makes (this) robot trustworthy?" Most of the people in today's society see robots as modern appliances or gadgets for not-everyday use. However, some of the people can be interested if such a machine would be safe to work with.

Answers for the question "what makes (this) robot trustworthy" can be interpreted that the design and human-like movement were the elements of a robot that made people more trusting the robot. However, there were some individuals that appreciated more the state-of-the-art interaction algorithms presented by the Pepper running interaction software implemented by the Weegree company.

The next graph, shown in Fig. 9, presents the answers to the question: what makes you feel uncomfortable in interacting with (the) robot. The reality and the experience of working with people and with humanoids give the following conclusion: the society (especially the oldest and the youngest parts of society) seem to be afraid of humanoid robots. Educators who work with both robots and children say that there are some incidents in which the little children may become a threat to robot – it has been also observed and analyzed in literature

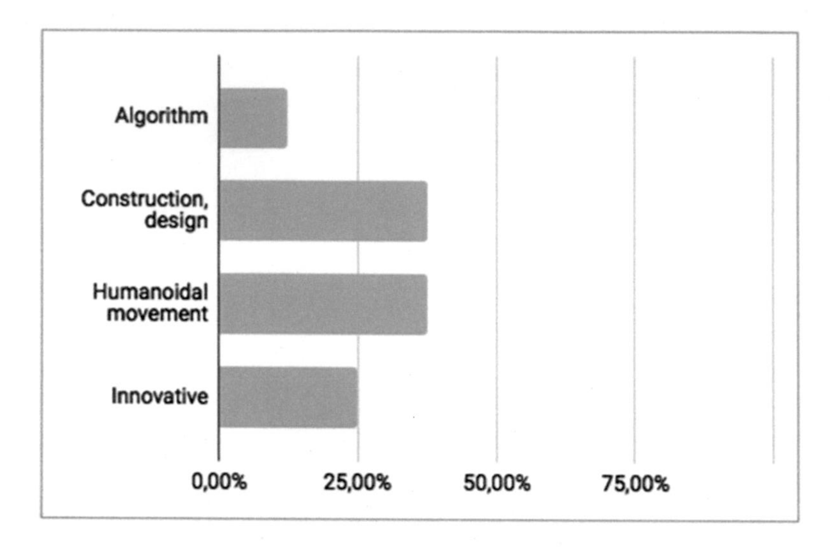

Fig. 8. Answers for the question "what makes (this) robot trustworthy?"

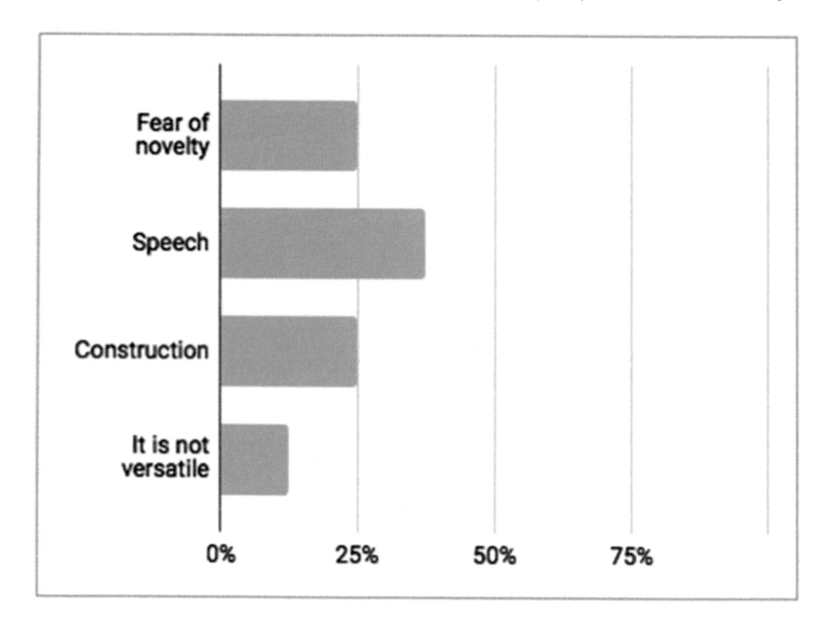

Fig. 9. What makes you feel uncomfortable in interacting with (the) robot?

[7]. Some adults though feel uncomfortable regarding the construction of the robot, or its way of generating speech. Those factors may be caused by the Uncanny Valley effect – on both sides of the valley – either too robotic or too real.

As seen in the graph above (Fig. 9), people gave (only) four answers to the question. The smallest group of the interlocutors said that the robot is not versatile, or universal. People said that this made them a little bit confused and uncomfortable to work with a robot, because Pepper – for example – is not able to hold anything heavy in its hands properly. It limits the usage of the robot and people have noticed that.

Two answers were said almost by the same number of people: "fear of novelty" and "construction of a robot". As authors said before, modern technology, or to put it in another way, novelty may be a reason for some people to feel uncomfortable when dealing with a robot. When it comes to the "construction" answer – humanoid movement, gesticulation, simulated facial expressions, etc. is in a way similar for the people when they say they feel uncomfortable in interactions with a robot.

The reason of the most uncomfortable situation when interacting with a robot was speech. Many people were not familiar with how Pepper was talking – high-pitched, robotic and loud voice influenced the overall user experience.

The last chart, presented in Fig. 10, reveals the answers to the key question: "does the eye contact make you feel uncomfortable?". It should be noticed, that especially for those interlocutors who think that the modern technology makes them feel uneasy and uncomfortable, asking question "does the eye contact make you feel uncomfortable" might imply the suggestion for the answer for it, such as yes or probably yes.

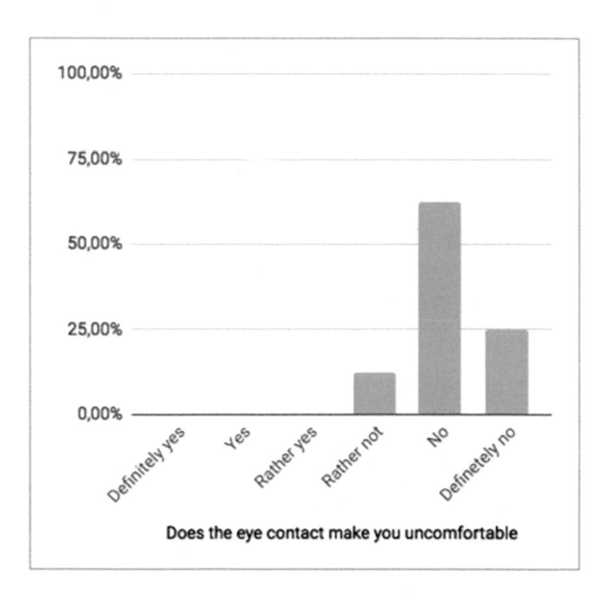

Fig. 10. Graph shows answers for a question: does the eye contact make you uncomfortable?

The authors' main research goal was to determine the reasons and population parameters of those individuals that have lower user experience due to the robot's eye contact. However, there was not a single answer to fit this view. In the Fig. 10 it can clearly be seen, that all of the answers were more or less disagreeing with the "suggested" answer. Even though most of the people were not fully comfortable with the construction or speech, such an aspect as the eye contact was not fundamental in making tester feel uncomfortable.

Overall conclusions of the graphs above are discussed in chapter conclusion and future work.

5 Conclusion and Future Work

In this paper authors have described significant aspects of user experience when interacting with a robot as well as its capabilities and advantages. Testing procedure, application and questionnaire were developed and discussed.

In the chapter Procedure and quantitative study authors included graphs presenting the data acquired during test procedure. It can be clearly seen (Fig. 5.), that the eye contact duration time may vary from person to person, but most of the people looked into Pepper's eyes between three and five seconds.

There were few questions asked in questionnaire before the final question (if eye contact makes them uncomfortable) was asked. It was done so for better conclusion to give as well as to get the background information before a person notices the point of the research. Answers for question about first impression (Fig. 6) show that the majority of people were impressed with modern technology and were fascinated about this. These answers show that young people look at the technology of modern devices, and that they are not overwhelmed by the functionality or behavior of the robot.

Figure 7 presents answers to the question if the robot would be useful as a companion. Most of the people would like to have a companion such as Pepper.

Another two graphs (Figs. 8 and 9) present answers to questions: "what makes robot trustworthy" and "what makes you feel uncomfortable". Unexpectedly, most of the answers for the first question were about humanoid movement capabilities and its design, and regarding the second question: speech and voice pitch were the main factors making testers feel uncomfortable. Because the age range of testers was between 24 to 56 years old, so those people are expected to be used to modern technology. In future studies, age range of people will be significantly expanded.

When asked if eye contact makes them uncomfortable, all the answers were "no" or "probably not". When authors were completing the questionnaires together with testers during interviews, most of those people said, that eye contact is not disturbing them at all. Furthermore, testers said that the eye contact is crucial to be able to fully experience human-like level of interaction, and that it significantly influences overall user experience of the robot.

Concluding, the eye contact was not making people feel uncomfortable or anxious in any way.

Future work will relate to human-robot interaction, including comparative study of human reactions to various AI engines installed on the same robotic platform, as well as non-verbal communication reinforcement by using context-aware gestures.

References

1. Needham, J.: History of Scientific Thought. Science and Civilisation in China, vol. 2. Cambridge University Press, Cambridge (1991). ISBN 0-521-05800-7
2. Byford, S.: SoftBank announces emotional robots to staff its stores and watch your baby Pepper will go on sale for under $2,000 in February. Vox Media, 5 June 2014. http://theverge.com. Accessed 10 June 2017
3. SoftBank Robotics: Find out more about Pepper (2015). http://www.ald.softbankrobotics.com/en/cool-robots/pepper/find-out-more-about-pepper. Accessed 11 Apr 2017
4. Shea, M.: Pepper: A robot to watch you sleep. The Skinny, 23 September 2015. http://www.theskinny.co.uk/tech/features/pepper-robot. Accessed 30 June 2017
5. SoftBank Robotics: Aldebaran 2.5 Documentation - Pepper robot (2016). http://doc.aldebaran.com/2-5/home_pepper.html. Accessed 01 Mar 2017
6. SoftBank Robotics: ALGazeAnalysis API Documentation (2014). http://doc.aldebaran.com/2-5/naoqi/peopleperception/algazeanalysis-api.html. Accessed 20 May 2017
7. Nomura, T., et al.: Why do children abuse robots? In: 2015 ACM/IEEE International Conference on Human-Robot Interaction. http://www.rikou.ryukoku.ac.jp/~nomura/docs/CRB_HRI2015LBR2.pdf. Accessed 30 June 2017

Pole-Free vs. Minimum-Norm Right Inverse in Design of Minimum-Energy Perfect Control for Nonsquare State-Space Systems

Marek Krok(✉) and Wojciech P. Hunek

Department of Electrical, Control and Computer Engineering, Opole University of Technology, Opole, Poland
mar.krok@doktorant.po.edu.pl, w.hunek@po.opole.pl

Abstract. In this paper a comparison of energy cost for different types of perfect control structures is presented. It is shown, that there is a possibility to improve the said control strategy for nonsquare LTI MIMO discrete-time state-space systems in terms of robustness through seeking of minimum control energy. It is remarkable, that simulation examples made in Matlab/Simulink environment confirm the potential of pole-free control design not only in the context of maximum-speed and maximum-accuracy properties, but, what is important, in terms of minimum energy behavior.

Keywords: Perfect control · Minimum-energy · Pole-free approach
Parameter matrix inverses · State space · Robust control design

1 Introduction

The energy of control signals is, right after the speed and steady-error, one of most frequently used quality criteria for different control strategies. As the pole-free perfect control [1] has just solved the maximum-speed/maximum-accuracy problem, the minimum control energy seeking is still open issue worth of the future scientific research. In the classic perfect control design the minimum-norm right T-inverse has been used as its properties has been corresponded with the minimum-energy of control runs [2–5].

Despite all possible advantages of pole-free perfect control manner, the structural stability still should be highlighted. The classical minimum-norm right T-inverse in some cases could result in unstable control signals while the newly presented pole-free method always guarantees stability for all systems with stable transmission zeros, if any [6–10]. Nevertheless, for all already presented simulation examples the pole-free perfect control energy was incomparably higher than in cases with stable non-zero poles of the closed-loop plants.

A recently published study of interesting facts about minimum energy of perfect control [6] give rise to undertake the deeper scientific research concerning the pole-free version of said control strategy [1]. While the pole-free perfect

W. P. Hunek and S. Paszkiel (Eds.): BCI 2018, AISC 720, pp. 184–194, 2018.
https://doi.org/10.1007/978-3-319-75025-5_17

control is remarkably faster in terms of achieving the steady-state of interior signals, the relation between energy cost in different scheme scenarios is rather still unpredictable.

Now, having two equivalent perfect control designs it is right place to undertake effort of comparing them through the prism of stability and energy, oscillations and amplitudes of input signals in a number of simulation examples.

The paper is organized as follows. At the beginning, the description of considered systems is given in Sect. 2. In this paper it will be the classic state-space framework. In Unit 3 the perfect control algorithm for nonsquare LTI MIMO plants is reminded. In Sect. 4 the perfect control design implementing different types of inverses will be presented. The simulation examples in Unit 5 show crucial runs obtained in Matlab/Simulink environment. In the penultimate section the open problems will be briefed. The final conclusions will be given in last section of this paper.

2 System Description

Let us consider the nonsquare LTI MIMO discrete-time state-space systems described in the following manner

$$\begin{aligned} \mathbf{x}(k+1) &= \mathbf{A}\mathbf{x}(k) + \mathbf{B}\mathbf{u}(k), \quad \mathbf{x}(0) = \mathbf{x}_0, \\ \mathbf{y}(k) &= \mathbf{C}\mathbf{x}(k), \end{aligned} \tag{1}$$

with corresponding matrices of dimensions $\mathbf{A} \in \Re^{n \times n}, \mathbf{B} \in \Re^{n \times n_u}$, $\mathbf{C} \in \Re^{n_y \times n}$, where n, n_u and n_y are the numbers of state, input and output variables, respectively. Additionally, $\mathbf{x}(k)$ is the state vector in discrete time k whilst $\mathbf{x}_0$ denotes its initial condition. Due to well-known fact, that the perfect control is achievable only for systems with more input then output signals, we will only refer to right-invertible systems. The reachability of perfect control entails the last limitation that the parameter matrix $\mathbf{CB}$ need to be of full rank. At the very end we also need to assume, that the pairs $(\mathbf{A}, \mathbf{C})$ and $(\mathbf{A}, \mathbf{B})$ are observable and detectable, respectively.

3 Perfect Control for Nonsquare Systems

The perfect control algorithm is one of the Inverse Model Control (IMC). The most characteristic property of the mentioned control strategy is fact, that due to using the performance index in form of

$$\arg\min_{\mathbf{u}(k)} \{[\mathbf{y}(k+1) - \mathbf{y}_{\text{ref}}(k+1)]^{\text{T}}[\mathbf{y}(k+1) - \mathbf{y}_{\text{ref}}(k+1)]\}, \tag{2}$$

the output reaches its reference value right after time delay deriving from the system description. For classic state-space framework with single unit delay the perfect control law can be obtained by equating the one-step output predictor $\mathbf{y}(k+1) = \mathbf{CAx}(k) + \mathbf{CBu}(k)$ to the reference/setpoint $\mathbf{y}_{\mathrm{ref}}(k+1)$ as follows

$$\mathbf{CAx}(k) + \mathbf{CBu}(k) = \mathbf{y}_{\mathrm{ref}}(k+1), \tag{3}$$

finally to obtain

$$\mathbf{u}(k) = (\mathbf{CB})^{\mathrm{R}}[\mathbf{y}_{\mathrm{ref}}(k+1) - \mathbf{CAx}(k)], \tag{4}$$

where $(\mathbf{CB})^{\mathrm{R}}$ denotes any right-inverse of matrix product $\mathbf{CB}$. The different types of inverses has been widely presented in plethora of literature, see e.g. Refs. [1,11–13]. Now, for zero reference value the above-mentioned algorithm reducing to the perfect control law in form of

$$\mathbf{u}(k) = -(\mathbf{CB})^{\mathrm{R}}\mathbf{CAx}(k). \tag{5}$$

Note that the presented formula will be used to obtain the perfect control runs in simulation examples. It is crucial, that this equation clearly corresponds to the well-known state-feedback formula

$$\mathbf{u}(k) = -\mathbf{Kx}(k), \tag{6}$$

which will be useful during calculation of poles of closed-loop perfect control plants.

Remark 1. It is clear that in case of perfect control the linear state-feedback matrix $\mathbf{K}$ from Eq. (6) equals

$$\mathbf{K} = (\mathbf{CB})^{\mathrm{R}}\mathbf{CA}. \tag{7}$$

Now, the poles of system being under perfect control will now be calculated according to the following equation

$$det(z\mathbf{I} - \mathbf{A} + \mathbf{BK}) = 0, \tag{8}$$

where $\mathbf{I}_n$ denotes identity n-matrix. Of course, the calculation of poles of closed-loop system can be referred to the poles of plant itself with the following relation

$$det(z\mathbf{I} - \mathbf{A}^*) = 0, \tag{9}$$

where, naturally, the matrix $\mathbf{A}^*$ represents the closed-loop system as follows

$$\mathbf{A}^* = \mathbf{A} - \mathbf{BK}. \tag{10}$$

From now, the above presented matrix $\mathbf{A}^*$ will be used to the analysis of the state-feedback systems, i.e. perfect control systems, in context of stability, both in stability analysis and under simulation examples.

Knowing the basics of perfect control theory, let us now start, in the next section, the comparison between the pole-free and ‘minimum-norm’ perfect control designs.

4 Pole-Free vs. 'Minimum-Norm' Perfect Control Designs

The perfect control design has attracted considerable research interest due to the affinity with widely explored topic of matrix inverse calculation. In case of parameter matrices frequently used is minimum-norm right T-inverse in form of

$$(\mathbf{CB})_0^{\mathrm{R}} = (\mathbf{CB})^{\mathrm{T}}[\mathbf{CB}(\mathbf{CB})^{\mathrm{T}}]^{-1}, \tag{11}$$

which is unique one and there is no possibility to influence the parameters giving the inverse calculation result [7,14]. Due to the uniqueness there is a threat of unstable control signals depending of the nonminimum-phase property of controlled perfect control system. It is clear now, that the right T-inverse can be used as a comparison for all other inverses. On the other side, there is nonunique right σ-inverse described in the following manner

$$(\mathbf{CB})^{\mathrm{R}} = \beta^{\mathrm{T}}(\mathbf{CB}\beta^{\mathrm{T}})^{-1}, \tag{12}$$

where $\beta \in \Re^{n_y \times n_u}$ is so-called degree of freedom. The non-unique inverse can be used in particular to increase the robustness of said control strategy as a remedy for possible inconviniences deriving from both instability or later achieved steady-state [1,14,15].

Remark 2. Sometimes we refer to term of 'pole-free inverse'. Of course, the sentence should be understood in the context of all the poles of closed-loop perfect control are placed in zero during an application of specified right inverse.

For now it has been shown, that the pole-free scenario of perfect control algorithm results in faster achieved the steady-state of control signals, while for all perfect control cases the output remains at its reference. Moreover, for second-systems being under perfect control the sufficient condition of pole-free manner has already been presented in form of

$$\mathbf{A}^*(1,1) = -\mathbf{A}^*(2,2), \tag{13}$$

where the whole issue reduces to the process of selecting the proper degrees of freedom of right σ-inverse [1].

As the perfect control is remarkably faster under using the pole-free inverse there is a need to compare it with control scheme after application of the minimum-norm unique inverse in terms of this properties. Now, the energy cost of perfect control described in following manner

$$\mathrm{E} = \sum_{i=0}^{n_u} \sum_{k=0}^{K} u_i^2(k), \tag{14}$$

where K stands for a time horizon, will be used to compare the minimum norm and pole-free σ-inverse in the context of possibly low energy of perfect control input signals. The crucial simulation examples are included in the next section.

5 Simulation Examples

5.1 Example 1

Let us consider an LTI MIMO nonsquare plant described in state-space framework with corresponding matrices $\mathbf{A} = \begin{bmatrix} 0.4 & 0.11 \\ 0.21 & 0.37 \end{bmatrix}$, $\mathbf{B} = \begin{bmatrix} 0.17 & 0.04 \\ -0.2 & -0.3 \end{bmatrix}$, $\mathbf{C} = \begin{bmatrix} 0.9 & 0.3 \end{bmatrix}$ and $\mathbf{x}_0^{\mathrm{T}} = \begin{bmatrix} 9 & 4 \end{bmatrix}$. After using the minimum-norm inverse we obtain the poles of closed-loop perfect control system equal to $z_1 = 0$, $z_2 = 0.1813$ and energy of control runs $\mathrm{E}_{\text{s-p}} = 501.7899$. Plots of signals obtained in this scenario are shown in Fig. 1.

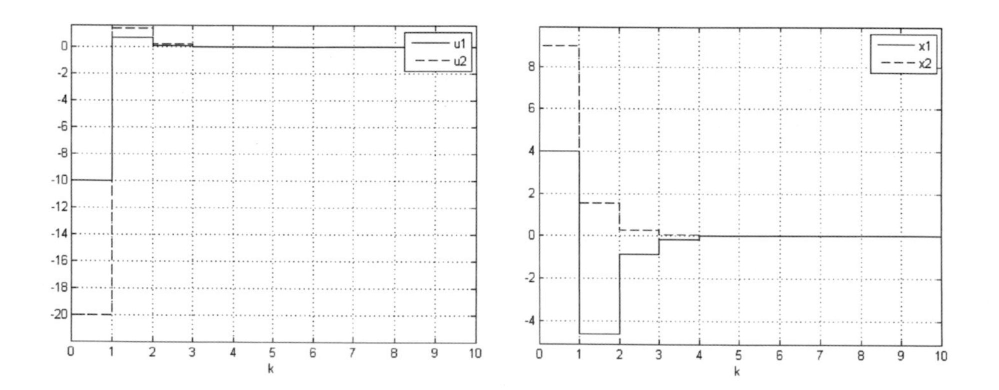

Fig. 1. Stable-pole perfect control, case: T-inverse

Now, under pole-free inverse case with proper selected $\beta = [-23.9808 \; 46.9484]$ we have

$$\mathbf{A}^* = \begin{bmatrix} -0.5430 & -0.1810 \\ 1.6291 & 0.5430 \end{bmatrix},$$

with all poles of closed-loop perfect control system equal to zero. It is not worth worrying, that in this case the control energy is higher than in 'minimum-norm' scenario and is equal to $\mathrm{E}_{\text{p-f}} = 1508.2478$. The maximum-speed/maximum accuracy runs obtained through an application of the pole-free perfect control design are shown in Fig. 2.

We also present the graph of output variable, valid for both stable-pole and pole-free cases (Fig. 3).

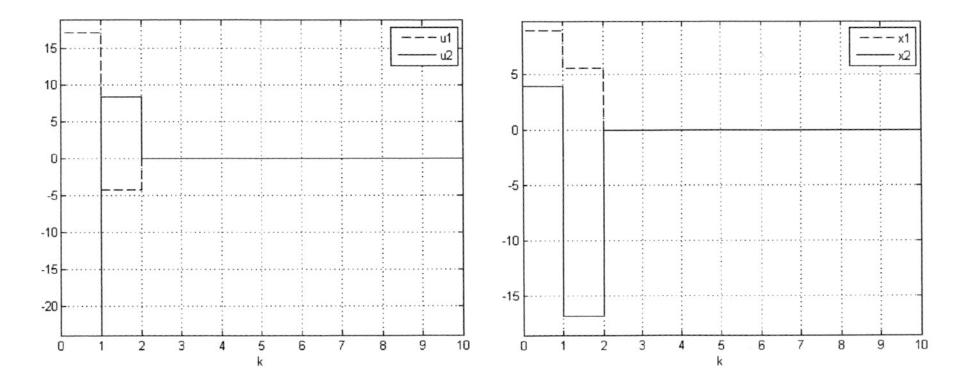

Fig. 2. Pole-free perfect control, case: σ-inverse

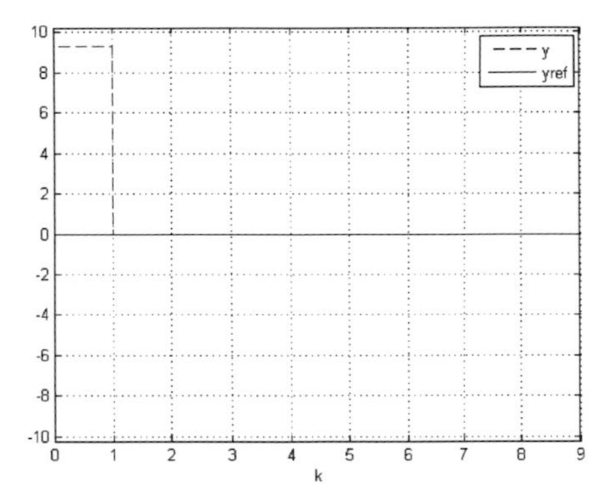

Fig. 3. Perfect control, both cases

5.2 Example 2

Let us consider, again, the LTI MIMO nonsquare plant described in the state-space framework with corresponding matrices $\mathbf{A} = \begin{bmatrix} 0.3 & -0.81 \\ 0.4 & 0.1 \end{bmatrix}$, $\mathbf{B} = \begin{bmatrix} 0.88 & 0.12 \\ -0.2 & 0.5 \end{bmatrix}$, $\mathbf{C} = \begin{bmatrix} 0.1 & 0.3 \end{bmatrix}$ and $\mathbf{x}_0^{\mathrm{T}} = \begin{bmatrix} 5 & -5 \end{bmatrix}$. Signals obtained during the simulation of the minimum-norm stable-pole perfect control system are shown in Fig. 4.

In this scenario the control energy is $\mathrm{E}_{\text{s-p}} = 54.6864$ while the poles are equal to $z_1 = 0$ and $z_2 = 0.2976$. In opposite, the pole-free runs are shown in Fig. 5.

The pole-free runs were obtained for $\beta = [7.6336 \ 13.8313]$, which clearly placed all eigenvalues of matrix

$$\mathbf{A}^* = \begin{bmatrix} 0.2112 & 0.6359 \\ -0.0707 & -0.2112 \end{bmatrix},$$

exactly at zero. In this case, the control energy is $E_{\text{p-f}} = 47.0366$. It is rather strange, that there is a solution, for which the energy cost is lower than in case of

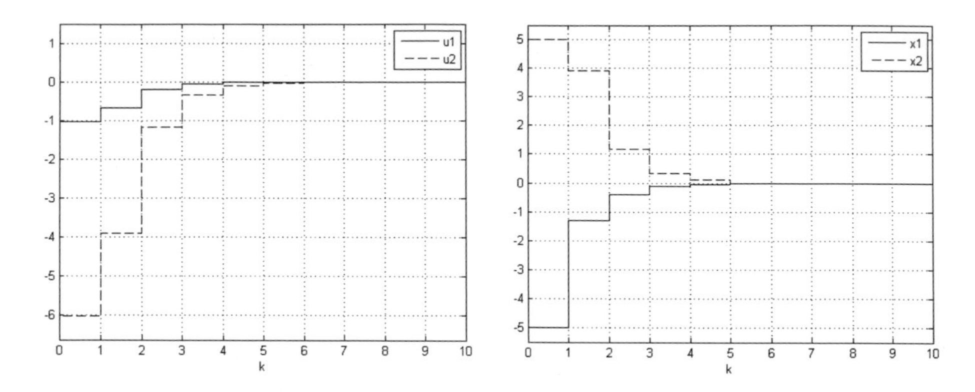

Fig. 4. Stable-pole perfect control, case: T-inverse

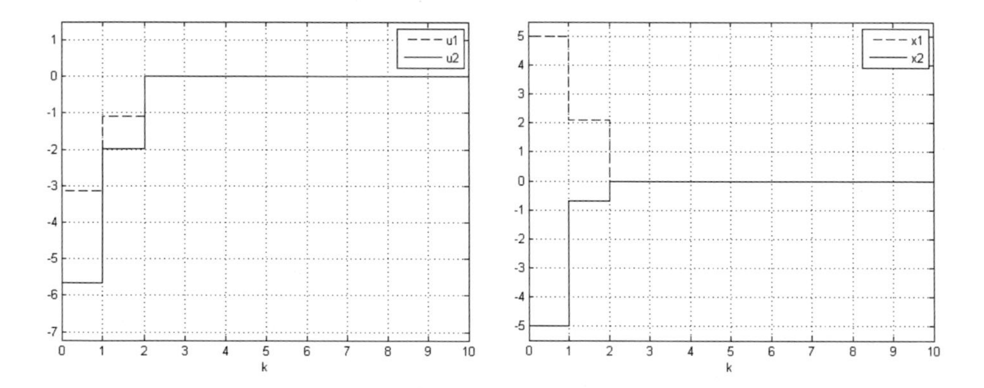

Fig. 5. Pole-free perfect control, case: σ-inverse

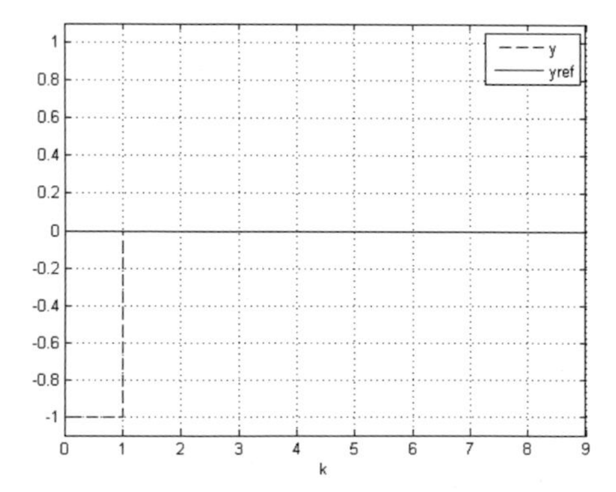

Fig. 6. Perfect control, both cases

previous stable-pole perfect control scenario. Especially noteworthy is the fact, that the pole-free perfect control was not only faster than the minimum-norm case, but also the zero-steady state is just reached under lower energy cost. Naturally, we present also the perfect control output in Fig. 6.

5.3 Example 3

Finally, let us consider an LTI MIMO nonsquare plant described in the state-space framework with corresponding matrices $\mathbf{A} = \begin{bmatrix} 0.4 & -0.7 \\ 1.1 & 0.2 \end{bmatrix}$, $\mathbf{B} = \begin{bmatrix} 0.4 & 0.8 \\ -0.1 & 0.4 \end{bmatrix}$, $\mathbf{C} = \begin{bmatrix} 0.2 & 0.4 \end{bmatrix}$ and $\mathbf{x}_0^{\mathrm{T}} = \begin{bmatrix} 7 & 5 \end{bmatrix}$. After using the minimum-norm right T-inverse we immediately arrive at poles equal to $z_1 = 0$ and $z_2 = -0.6885$. The control and state signals can be observed in Fig. 7. In this case the control energy is $\mathrm{E}_{\text{s-p}} = 599.2593$.

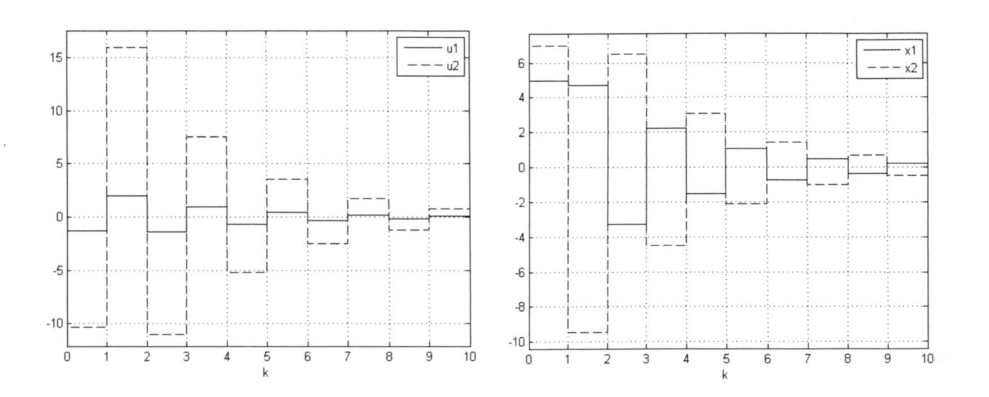

Fig. 7. Stable-pole perfect control, case: T-inverse

Again, after using the right σ-inverse with appropriate degree(s) of freedom $\beta = [5.2632 \ \ -5.0000]$ we obtain

$$\mathbf{A}^* = \begin{bmatrix} -0.3091 & 0.6182 \\ -0.1545 & -0.3091 \end{bmatrix}.$$

with all closed-loop perfect control poles equal to zero. In this pole-free scenario the control energy is $E_{\text{p-f}} = 532.5040$. The following runs of control and state signals are shown in Fig. 8.

Again, the energy needed to obtain the steady-state is much lower then the energy obtained in the minimum-norm scenario. Moreover, we can observe that in this pole-free case the minimum-norm inverse was outperformed in design of perfect control not in terms of speed and energy cost, but also in the context of the maximum amplitude of the control signals. At last, in Fig. 9 we present the output run received it the last simulation example.

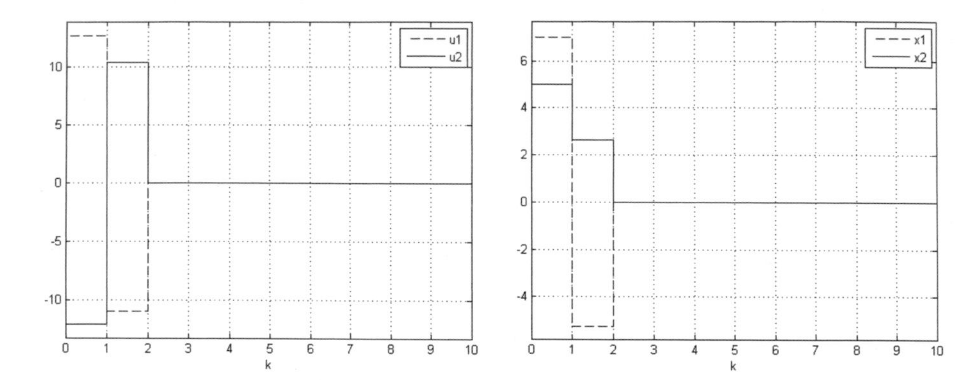

Fig. 8. Pole-free perfect control, case: σ-inverse

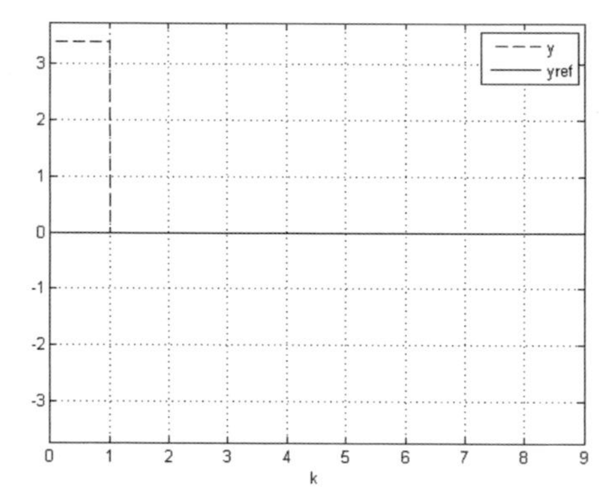

Fig. 9. Perfect control, both cases

6 Open Problems

6.1 The Global Minimum of Control Energy

Recently presented results show, that the minimum-norm right inverse do not ensure the minimum energy of the control signals for plants being under perfect control. This fact raises a question about existence of inverses, that will always result in the possible minimum control energy. There might be a need to introduce the performance index which will take into consideration both the minimum-error and minimum-energy requirements.

6.2 The Minimum State Energy

As the pole-free perfect control can compete with the minimum-norm stable-pole perfect control in sense of control energy, the energy of state variables still might be the challenging topic. So the issue is to find such inverse that will lead to the minimum energy of the state variables, naturally with respect to the minimum-error demand.

6.3 The Initial Condition Issue

It is possible, that the initial conditions play undeniable role in determining whether the minimum-norm or pole-free inverse will have lower control energy. The wide research study is still needed to explore in this field of control theory. As the energy cost is a result of transition between the initial and steady state under control action, not only the parameters of system matrices may be important. Nevertheless, coupling the parameters of system matrices with initial conditions is interesting research problem.

7 Summary

The comparison of two different perfect control designs has been presented in this paper. The pole-free perfect control obtained through the special selected degree(s) of freedom of right σ-inverse clearly outperforms in many terms the stable-pole one involving the unique right T-inverse. Notwithstanding, the simulation examples show, that the topic of minimum-energy perfect control is very complex problem, not to be simplified to one-sided study of different inverses. The properties of closed-loop perfect control under different design strategies seems to be rather unpredictable, so this topic still needs to be scientifically explored.

References

1. Hunek, W.P., Krok, M.: Pole-free perfect control for nonsquare LTI discrete-time systems with two state variables. In: Proceedings of the 13th IEEE International Conference on Control and Automation (ICCA 2017), Ohrid, Macedonia, pp. 329–334 (2017). https://doi.org/10.1109/ICCA.2017.8003082
2. Stanimirović, P.S., Petković, M.D.: Computing generalized inverse of polynomial matrices by interpolation. Appl. Math. Comput. **172**(1), 508–523 (2006). https://doi.org/10.1016/j.amc.2005.02.031
3. Karampetakis, N.P., Tzekis, P.: On the computation of the generalized inverse of a polynomial matrix. IMA J. Math. Control Inf. **18**(1), 83–97 (2001). https://doi.org/10.1093/imamci/18.1.83
4. Ben-Israel, A., Greville, T.N.E.: Generalized Inverses, Theory and Applications, 2nd edn. Springer, New York (2003)
5. Hunek, W.P.: Pole-free vs. stable-pole designs of minimum variance control for nonsquare LTI MIMO systems. Bull. Pol. Acad. Sci. Tech. Sci. **59**(2), 201–211 (2011). https://doi.org/10.2478/v10175-011-0025-y
6. Hunek, W.P.: New interesting facts about minimum-energy perfect control for LTI nonsquare state-space systems. In: Proceedings of the 22nd IEEE International Conference on Methods and Models in Automation and Robotics (MMAR 2017), Międzyzdroje, Poland, pp. 274–278 (2017)
7. Hunek, W.P.: Towards a General Theory of Control Zeros for LTI MIMO Systems. Opole University of Technology Press, Opole (2011)

8. Hunek, W.P., Latawiec, K.J., Stanisławski, R., Łukaniszyn, M., Dzierwa, P.: A new form of a σ-inverse for nonsquare polynomial matrices. In: Proceedings of the 18th IEEE International Conference on Methods and Models in Automation and Robotics (MMAR 2013), Międzyzdroje, Poland, pp. 282–286 (2013). https://doi.org/10.1109/MMAR.2013.6669920
9. Hunek, W.P., Latawiec, K.J., Majewski, P., Dzierwa, P.: An application of a new matrix inverse in stabilizing state-space perfect control of nonsquare LTI MIMO systems. In: Proceedings of the 19th IEEE International Conference on Methods and Models in Automation and Robotics (MMAR 2014), Międzyzdroje, Poland, pp. 451–455 (2014). https://doi.org/10.1109/MMAR.2014.6957396
10. Dadhich, S., Birk, W.: Analysis and control of extedned quadruple tank process. In: Proceedings of the 13th IEEE European Control Conference (ECC 2014), pp. 838–843 (2014). https://doi.org/10.1109/ECC.2014.6862290
11. Hunek, W.P.: An application of new polynomial matrix σ-inverse in minimum-energy design of robust minimum variance control for nonsquare LTI MIMO systems. In: Proceedings of the 8th IFAC Symposium on Robust Control Design (ROCOND 2015), Bratislava, Slovakia, pp. 150–154 (2015). https://doi.org/10.1016/j.ifacol.2015.09.449
12. Hunek, W.P.: New SVD-based matrix H-inverse vs. minimum-energy perfect control design for state-space LTI MIMO systems. In: Proceedings of the 20th IEEE International Conference on System Theory, Control and Computing (ICSTCC 2016), Sinaia, Romania, pp. 14–19 (2016). https://doi.org/10.1109/ICSTCC.2016.7790633
13. Hunek, W.P.: An application of polynomial matrix σ-inverse in minimum-energy state-space perfect control of nonsquare LTI MIMO systems. In: Proceedings of the 20th IEEE International Conference on Methods and Models in Automation and Robotics (MMAR 2015), Międzyzdroje, Poland, pp. 252–255 (2015). https://doi.org/10.1109/MMAR.2015.7283882
14. Levine, W.S. (ed.): The Control Handbook. Electrical Engineering Handbook. CRC Press and IEEE Press, Boca Raton (1996)
15. Grimble, M.J.: Controller performance benchmarking and tuning using generalised minimum variance control. Automatica **38**(12), 2111–2119 (2002). https://doi.org/10.1016/S0005-1098(02)00141-3

Meta-structural Graph-Based Design Patterns for Knowledge Representation in Association-Oriented Database Metamodel

Marcin Jodłowiec[1(✉)], Marek Krótkiewicz[1], and Krystian Wojtkiewicz[1,2]

[1] Faculty of Computer Science and Management, Wroclaw University of Technology, Wroclaw, Poland
{marcin.jodlowiec,marek.krotkiewicz,krystian.wojtkiewicz}@pwr.edu.pl
[2] Institute of Control Engineering, Opole University of Technology, Opole, Poland

Abstract. This paper describes the problems of modeling graph-based structures in Association-Oriented Database Metamodel in the context of knowledge representation system. The basics of Association-Oriented Metamodel solutions, principles of modeling and sample implementations of graph structures have been presented, including labeled graphs as well as generalization of graphs, i.e. hypergraphs. Subsequently, metastructural ontological design patterns dedicated to knowledge representation systems are presented based on example of standard class-instance-feature-value and relationship patterns.

1 Introduction

One of the fundamental issues that knowledge engineering focuses on is the problem of knowledge representation. It concentrates on building such data structures that are a structural basis for artificial intelligence systems. It also means the construction of such data models, which will be able to store the description of knowledge in an efficient way, including the least possible restrictions, and unequivocal, especially with regard to the possibility of its subsequent use in the inference process. The knowledge should be considered as data along with information on how to interpret it.

It is worth mentioning that the structures of data and knowledge representation do not have semantics by themselves. It is contained in the domain model of represented reality, e.g. in the form of ontology [13,21]. In information systems the semantics of data and relations between them is very often contained in a description provided in natural language, or it remains in the mind of the modeling engineer. This happens when the model expressiveness[1] in relation to complexity of modeled structures. In particular, this applies to very complex models, e.g. those that are the foundations of knowledge representation.

[1] By the term *expressiveness* authors understand the ability of language to express constructions and concepts of thought. It is a measure of phrases brevity in the language in relation to the complexity of the structure of thought, which it carries.

W. P. Hunek and S. Paszkiel (Eds.): BCI 2018, AISC 720, pp. 195–206, 2018.
https://doi.org/10.1007/978-3-319-75025-5_18

Modeling the structure of knowledge bases designers are faced with the need to resolve issues of conceptualization. Conceptualization is understood as a process of extracting the concepts and the relations between them from the domain or part of reality. It should be stressed that each concept shall be defined by relationships to other concepts. There may be any number of such relationships, they can have different meaning, as well as there may also be many participants taking part in them. In this approach, while modeling concept-oriented knowledge representation system, the relationship becomes basic semantic category. Therefore, a natural need for a model oriented on relationships, that allows a comprehensive approach arises. The relationships should be defined precisely to include as much semantics as possible. Due to that, structures taken from graph theory are commonly used, including simple graph, directed graph and hypergraphs [3,8,24] which are a graph structures generalization in terms of edge arit Graphs are an abstract tool i.e. they are separated from the implementation and specific use cases. This approach has long been used in the modeling of knowledge representation structures [14–16,18]. Lets consider the concept of the semantic networks, that in its basic form is nothing more than a graph, where the vertices represent concepts, while the edges represent the relationships between those concepts [1,12,20]. No restrictions given to the labeling of edges allows the representation of any relationships between concepts.

The paper is organized in the following order. The next section contains the research problem. Section 3 shows the basic properties of AODB crucial in terms of its use as data layer for graph-based knowledge representation structures, what is presented in Sects. 4 and 5. The native support for graphs and hypergraphs has been outlined there, including such features as possibility to create n-ary associations, free definition of multiplicity on both sides of roles as well as separation of data and relationships.

2 Modeling Problem

Graph theory provides a powerful theoretical tool for modeling knowledge representation structures. Graphs and hypergraphs in a natural way model the relations between objects appearing in a modeled reality. In graph the relationships are binary while the notion of hypergraph removes this limitation. However, the concept of graph is too general and abstract to use it directly as knowledge representation structure. It is a concept which constitutes an abstract, high-level perception model of data and data structures, and at the same time also perception of knowledge representation structures. Graph theory presents a value to knowledge representation. Such support is given only by models and tools that natively support n-ary relations, which is, among others, the metamodel described in the Sect. 3 of the paper.

The paper provides a discussion on the usage of graph structures in knowledge representation, that is not limited to semantic networks. In authors' opinion most of the nowadays approaches are inevitably based on graph structures. The main research problem was to find a data layer flexible enough to support such structures. The Association-Oriented Database Metamodel (AODB) [10] is presented as lingual platform for knowledge representation structure patterns based on the hypergraphs. This metamodel natively support e.g. n-ary relations, what makes it a perfect choice as data layer for knowledge representation systems such as Semantic Knowledge Base (SKB) [12]. However due to the fact, SKB is outside of paper's scope, the expressiveness of presented database metamodel is shown through basic graph and hypergraph structures. It should be noted that the design patterns presented in this paper have meta-structural characteristics. This means that they are a description of how to create data structures. Thus, they set the framework for the creation of graphs, which will represent a specific occurrence (instance) of created meta-structures. Following, the support given by the use of graphs to the implementation of the fundamental ontological properties is described. Let us consider the meta-structure representing relationship of *part-whole* or *gen-spec*, and their instances will be specific relations that describe relationships of a *part-whole* or *gen-spec* e.g. *car-engine* or *vehicle-car*.

2.1 Related Work

The problem of the representation of ontological knowledge is known in information sciences since 1960s [23]. There are number of solutions elaborated, as well as many ways to specify ontological knowledge. The survey in regard to languages used in ontology definition has been very well presented in [2,4]. Nowadays, a large emphasis is put on the development of Semantic Web [17,22]. The majority of proposed solutions concern primarily on the idea of a semantic web based on Web Ontology Language (OWL) which is built upon the standard of Resource Description Framework (RDF). and form a separate branch of systems based on ontological knowledge. There are also other autonomous languages of the ontology definition and independent projects, such as KIF [7], OKBC [5], CycL [6] or Conceptual graphs [19].

3 Main Features of Association-Oriented Database Metamodel

Association-Oriented Database Metamodel (AODB) is a research project of database metamodel comprising various elements. All of its components have been designed from the beginning to the end in a way dedicated for this metamodel only, i.e. it does not use any languages, models, data storage solutions or any other elements of known database systems. Association-Oriented Database Metamodel is based on the following basic primitives [9,10]:

- intensional (structures): *database* (*Db*), *association* (*Assoc*), *collection* (*Coll*), *role* (*Role*), *attribute* (*Attr*),
- extensional (data): *association object* (*AssocObj*), *object* (*Obj*), *role object* (*RoleObj*).

In the intensional sense two the most important primitives are: *association* and *collection*. *Association* is the primary concept that implements the relationships of data, while *collection* is responsible for data storage. *Association* owns *roles*. *Roles* are lists of references to connected elements. The elements connected may be either objects (instances of *collection*) and *association objects* (instances of *association*). The number of *roles* in the *association* is not restricted, thus the relationships can be of any arity. Each role is determined by a number of properties such as uniqueness within the *association* identifier (name) of role, multiplicity on the side of *association*, multiplicity on the side of the related element, lifetime dependency of related elements by relating element, as well as lifetime dependency of relating element by related element, moreover navigability and reduction on the number of repetitions of related elements.

In addition to standard roles in AODB there is defined auxiliary type of binding, namely *association description*. In a sense, and with some approximation it can be treated as a specific and predefined type of role, characterized by the fact that it does not have its name, can occur only between the association and the collection, is unidirectional i.e. *describing objects* have no information about this fact, multiplicity on the side of the collection is 0..1 and there is no possibility of mutual limitation of lifetimes of any of linked elements.

Collection corresponds to the concept of data storage. In the intensional sense *collection* it is determined by a set of attributes that define the types of values stored in objects. *Association* does not have the ability to store data, and *collection* does not have the ability to create relationships, because the internal structure of these Association-Oriented Metamodel elements forces completely unique use of them. Both of these categories are independently subject to the mechanisms like inheritance. *Association* can inherit from other *associations* and in the same way there can be created inheritance trees of *collections*. The separation of data and relations is complete also due to the fact that in AODB each primitive is completely distinct and performs only one function. That means that if the database scheme will have any *collection* by definition of AODB they perform a one and only one precise function in the grammatical sense. There is no situation known from the relational metamodel or object model (OM ODMG 3.0), where *relation* or respectively *class* can perform data storage function (tuples or objects) along with implementing n-ary relationships. AODB is completely unambiguous in the sense of the function of individual grammar elements of metamodel. This property is very important not only in modeling stage but also during the analysis of the existing model or in case of modification of complex database schemas.

Apart from the formal definition of *Association-Oriented Database Metamodel* ($\mathfrak{M}$) within AODB there was developed descriptive part ($\mathfrak{D}$) and behavioral part ($\mathfrak{B}$). Descriptive part contains *Formal Notation* (AFN), which is strict, concise, formal and symbolic language to describe both intensional and extensional parts of *association-oriented database*. *Association-Oriented Modeling Language* (AML) is a graphical language of structures and data in AODB. It is fully consistent with the AFN and sole definition of metamodel. Both description languages i.e. AFN and AML are completely dedicated to AODB and they do not constitute a modification or a subset of any existing language. Behavioral part of AODB includes *Association-Oriented Query Language* (AQL) and *Association-Oriented Data Language* (ADL). These languages are corresponding fully to the metamodel and are completely original solution to the problem of data extraction and modification since they act directly on the hypergraph structures underlying AODB data model.

4 Graph Structures in Association-Oriented Database Metamodel

The graph modelling problem solutions prepared in AODB are shown in Figs. 1, 2 and 3. The presented diagrams are designed in AML, a dedicated modeling language for database schemas in AODB. *Associations* in AML are represented by diamonds, while *collections* by rectangles with three separate areas. The solid lines represent *role* with a small circle located at *association*, which is the owner of the role. *Describing collections* LABEL are linked with *associations* Edge by the *description* compound type indicated by a dotted line.

The first model of a AODB structure representing graph located in the Fig. 1 shows an example where the nodes are represented as *association* Node with the *describing collection* NODE that stores additional data referring to the node (name, value). The edges have been modeled in the form of *role* Edge belonging to *association* Node. In this way, each node has information about its attributes (*collection* NODE) and the outgoing edges. This is the easiest way to model graph structure in the AODB.

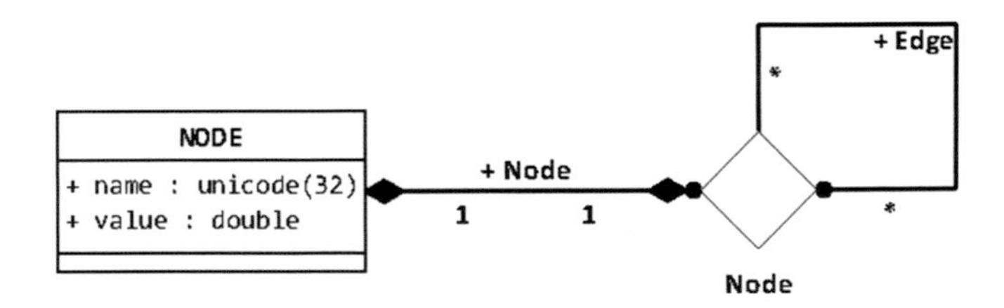

Fig. 1. Role-based implementation of simple graph in AODB

The second model (Fig. 2) shows a graph structure in which each edge belongs to *association* Edge, and each node belongs to *collection* NODE. Edges and nodes are connected by *role* Node, which has multiplicity equal to exactly 2 on the side of the node. The multiplicity on the side of edges is not limited, as each node may be associated with any number of edges.

The third model in the Fig. 3 shows the almost the same idea with just a small addition of *describing collection* added to *association* Edge. It performs the functionality of the edge labeling. In this way, it can easily be extended from graph to network structures.

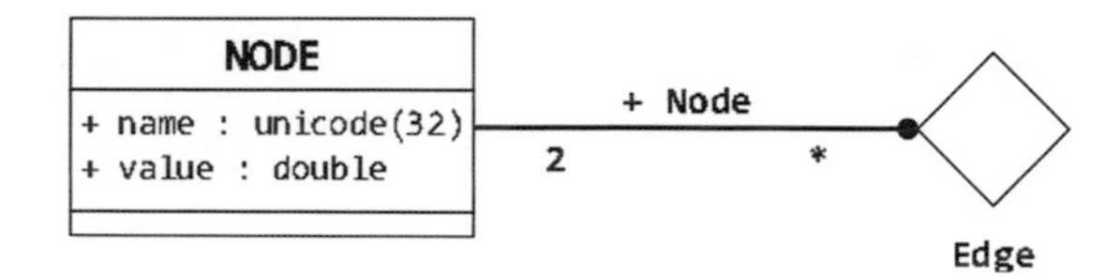

Fig. 2. Association-based implementation of simple graph in AODB

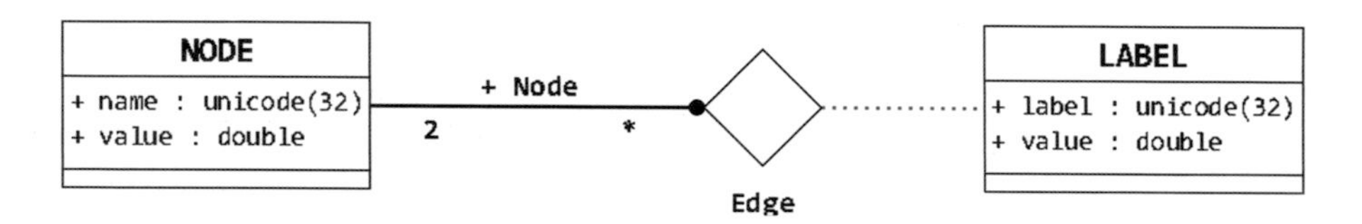

Fig. 3. Association-based implementation of labeled graph in AODB

The following association-oriented models (Figs. 4 and 5) represent hypergraph models that differ from the models in the Figs. 2 and 3 only by the multiplicity of *role* Node on the side of *collection* NODE. This provides the ability to connect multiple nodes to a single edge, which fulfills the conditions for the formation of hypergraph.

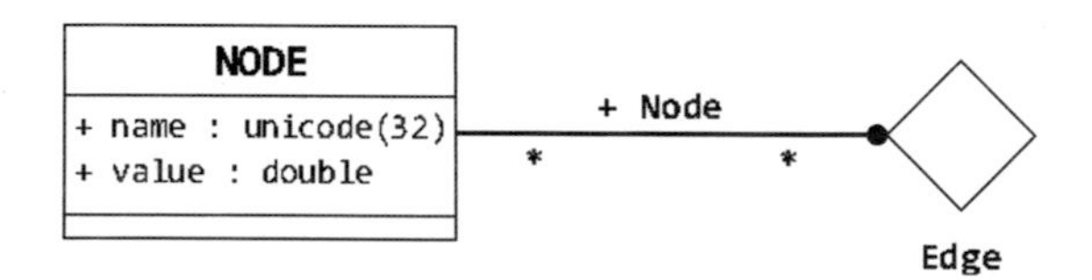

Fig. 4. Simple hypergraph implementation in AODB

Models presented in this section are the basic physical design patterns implementing the functionality of the graphs and hypergraphs in the Association-Oriented Database Metamodel. These structures are extremely simple, efficient, unambiguous and natural for presented data metamodel. These are but a starting point to create more complex structures of a meta structural and ontological characteristic.

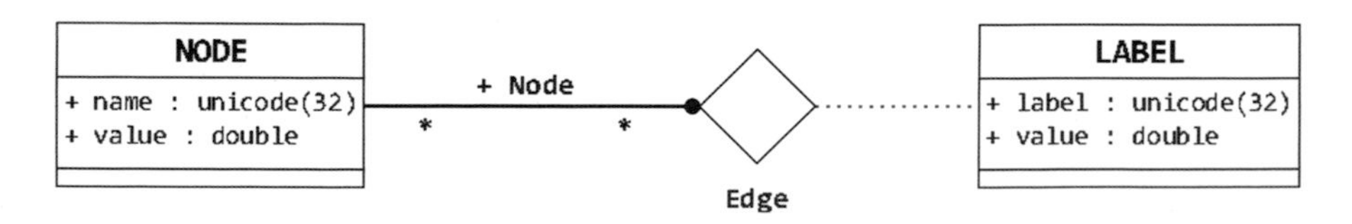

Fig. 5. Labeled hypergraph implementation in AODB

5 Implementation of Graph-Based Design Patterns for Knowledge Representation

5.1 Classes, Instances, Attributes, Values, Properties

The Fig. 6 introduces the design pattern for knowledge representation in the area of basic ontological information. The model comprises *classes* and *instances*, *features values* and *properties* thus the pattern will be referred as CIFVP.

Class is modeled as an *association* Class with describing it *collection* CLASS. A class can have multiple instances (*role* Instance). Instance can be assigned to at most one class. Classes also have features (*collection* FEATURE). Properties are defined as relationship between instance, attribute and values. It should be interpreted as for a given instance specified feature takes a value or set of values. Both *class*, *instance*, *feature* and *value* represent specific types of concepts. For this reason a generalization relationship has been defined (*IS-A*) between the collection CONCEPT and collections corresponding to aforementioned categories. Generalization relationship ensures both the inheritance of attributes and roles, rights to perform these roles, but also the inheritance of type. The latter property of generalization in AODB provides a substitution type in the behavioral context i.e. in the case of query to a specific node in a formula of dedicated query language.

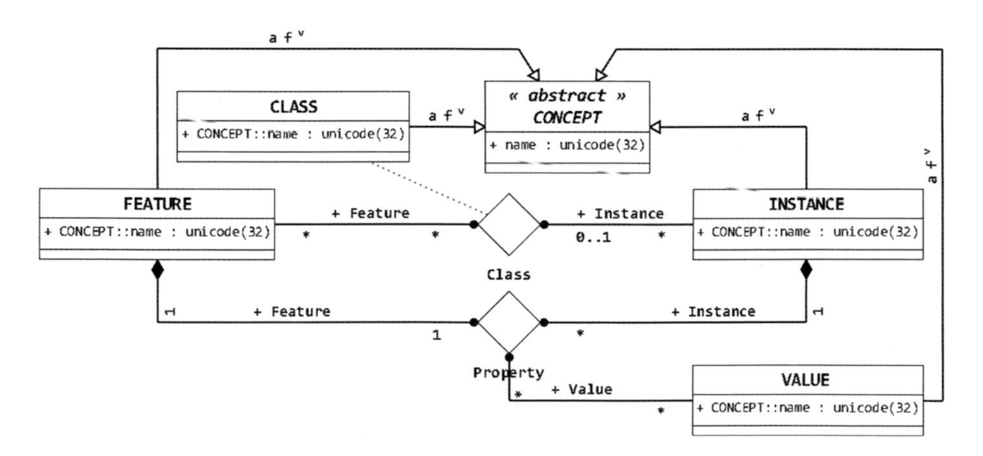

Fig. 6. The CIVFP design pattern comprising ontological knowledge representation in AODB: classes, instances, features, values and properties

Let us consider a class *car*, which will be represented in the system by appropriately collection object CLASS and association object Class. This class may have *features* such as *brand*, *model*, *registration number* what will be reflected by the appropriate objects of class FEATURE. *Association* Property assigns data values to the characteristics of each instance.

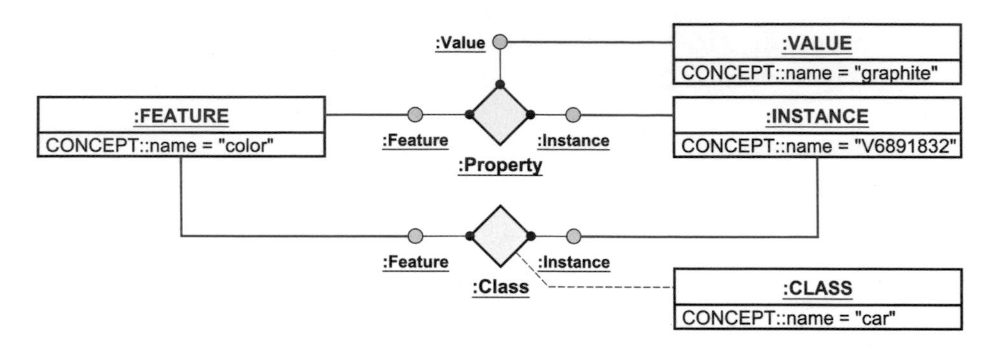

Fig. 7. Sample ontological knowledge data defined within the CIFVP design pattern

In terms of extensional part (Fig. 7), the class *car* appears as the *object* of *collection* CLASS with the value of attribute CONCEPT::name = *"car"*. This *object* describes exactly one *association object* of *association Class*, whilst the *association object* groups all instances of cars within his Instance *role*. A specific car instance (identified by *V6891832*) is the *object* of INSTANCE collection. The *association object* of Property *association* connects the car instance *object* with the object representing the feature and value (*objects* of FEATURE and VALUE *collections*).

5.2 The Types of Relationships and Their Instances

A model presenting the design pattern for knowledge representation in the field of relationships between concept is presented in Fig. 8. The RR (*Relationship-Role*) design pattern is an extension of graph idea, enriching the model with the ability to define types of relationships and their instances. Thus, data is stored on two levels: (1) meta-level (structure definition) and (2) the actual data. *Associations* Relationship and Role, together with the corresponding *collections* describing them, perform the task of the definition of relationships types (first level). The *associations* RelationshipInstance and RoleInstance are the second level of the pattern. The *role* Concept in *association* RoleInstance enables to materialize the relationship by linking it to actual concepts in knowledge base. The relationship *part-whole*, for example, can be defined as *association object* Relationship described by *object* RELATIONSHIP with the attribute *name* = *"part-whole"*. *Association object* created this way will have a list of two *association objects* Role, where each of them will be accordingly *part* and *whole*.

Relationship defined as such may be used repeatedly to create its instances, which provides *association* RelationshipInstance and *association* RoleInstance. *Association* RoleInstance has a list of concepts represented by *collection* CONCEPT. It corresponds to the specific concepts that can serve as *part* or *whole*. It should be noted that in AODB it is possible to connect directly *associations* with each other, which significantly increases the expressiveness of the meta-

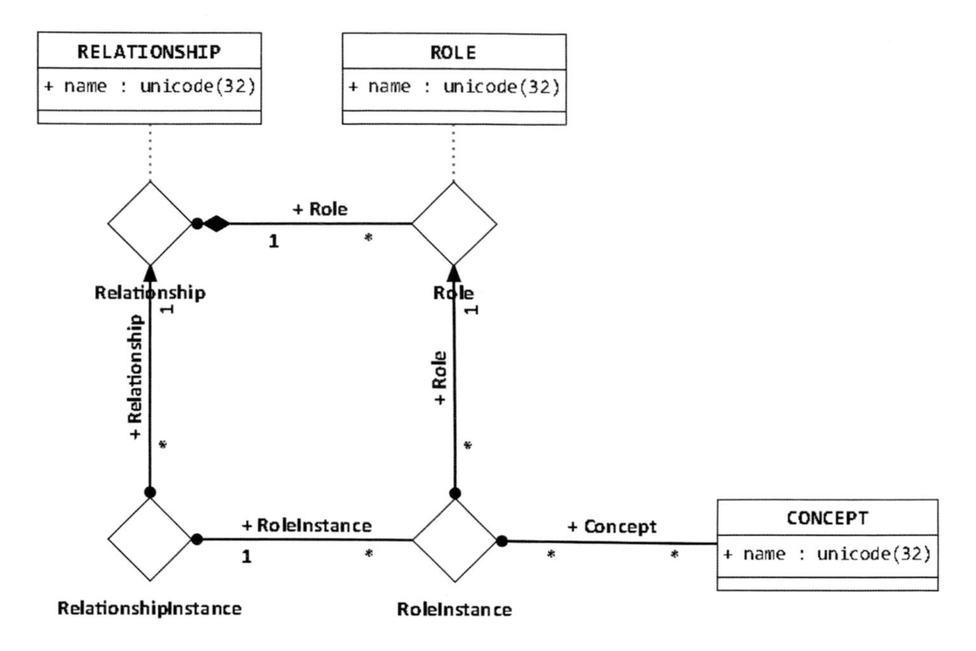

Fig. 8. The RR design pattern comprising ontological knowledge representation in AODB: relationship types and its instances

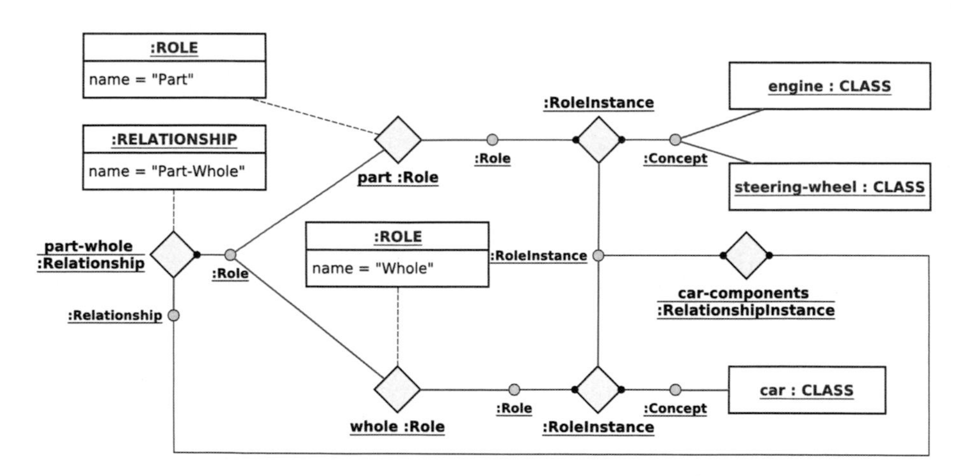

Fig. 9. Sample ontological knowledge data defined within the RR design pattern

model, and contributes significantly to its brevity and unambiguity of models being described.

Considering the extensional data model shown in the Fig. 9, it exemplifies some knowledge on the *part-whole* relationship regarding car structure, namely that car has the following parts: engine and steering wheel. The meta-level of the pattern defines an *association object* of *part-whole* relationship linked with two roles, which are represented by the *association objects* of *association* Role: *part* and *whole* (and corresponding describing objects). The data level of this model contains of the instantiation of the relationship in the form of *association object car-components* of *association* RelationshipInstance. Consequently, there are two respective instances of *association* RoleInstance. They are linked to *objects* representing *parts* (engine and steering wheel) and *whole* (car) concepts.

6 Summary

The paper presents examples of knowledge representation structures based on graphs, hypergraphs and ontologies. They should be treated only as a general description of the concept of knowledge modeling using AODB. The main objective was to propose structures of knowledge representation. The structures were modeled in Association-Oriented Database Metamodel and presented in Association Modeling Language. In contrast to e.g. UML the proposed solution is not purely conceptual, i.e. it does not require the implementation in a separate database system, since AODB is just such a system. The modeled structures are the physical form of the database, where users can perform modification operations (ADL) as well as data extraction operations (AQL). In addition AODB has grammatical constructions impossible to express in e.g. UML or major database metamodels such as relational model or object model[2] (e.g. n-ary relationships that directly implement hypergraph structures or constraining cardinality of elements in the relationship).

Presented models, though extremely simple and general, put special emphasis on the issues of free meta structures definition and their specific instances. This is particularly important when the system is built to provide e.g. full freedom in definition of relationships between concepts (RR design pattern, Fig. 8). The free relationships defining enables unambiguous and natural modeling of knowledge without the need to seek intermediate solutions or compromises. Figure 6 presents the CIFVP, the model that provides a the combination of the concept of representation *class-instance* and at the same time takes into account commonly used triple: *object–attribute–value*. Despite its simplicity, this model is completely unambiguous and in a very natural way combines extremely common descriptions of reality.

7 Conclusions

The proposed method of knowledge representation was implemented in AODB, which is its database layer. These examples demonstrate the high expressiveness, and also the unambiguity and naturalness of modeling. That naturalness of modeling results, among other things, in support of n-ary relationships and

[2] The standard of objects databases is OM ODMG 3.0.

in possibility of direct relationships connecting. One of the most important features of AODB is very clear, based on primitives definition, separation of data and relationships. This is particularly important for modeling the knowledge representation structures. It provides not only equivalence but also the ease of interpretation of the modeled structures. Semantics of models is derived in a very large extent directly from their grammar, and hence there is no problem of ambiguity in terms of functions performed by the various primitives, such as: *association*, *collection*, *role*, *attribute*, *value*, etc.

AODB enriches the available knowledge representation methods. It has been designed for complex systems capable of storing knowledge represented by a variety of methods, such as semantic networks, ontologies, graphs, and object-oriented or frame approach. The assumptions made before the designing of AODB focused mainly in the area of unambiguity and productivity in order to provide direct support to systems of knowledge representation in an efficient way, while natural for modeling at the same time.

Currently AODB is used to design the knowledge base system known as Semantic Knowledge Base (SKB) [11,12]. It is based on the knowledge representation structure that is fully implemented in AODB. Presented modeling concepts are directly reflected in the specific modules SKB, however, a thorough description of SKB is outside of the paper's scope. The main aim of the paper was to present alternative way of modeling knowledge representation structures based on formally defined, fully complementary and consistent metamodel dedicated to applications in the area of broadly defined knowledge engineering.

References

1. Collins, A.M., Quillian, M.R.: Retrieval time from semantic memory. J. Verbal Learn. Verbal Behav. **8**, 240–247 (1969)
2. Corcho, O., Gómez-Pérez, A.: A roadmap to ontology specification languages. In: International Conference on Knowledge Engineering and Knowledge Management, pp. 80–96 (2000). http://www.springerlink.com/index/QV8H33HQYB643Y14.pdf
3. Dipert, R.R.: The mathematical structure of the world: the world as graph. J. Philos. **94**(7), 329–358 (1997)
4. Djedidi, R., Aufaure, M.A.: Ontology evolution: state of the art and future directions. In: Ontology Theory, Management and Design: Advanced Tools and Models, p. 179 (2010)
5. Fikes, R., Karp, P.D., Rice, J.P.: OKBC: a programmatic foundation for knowledge base interoperability. In: Proceedings of the National Conference on Artificial Intelligence, pp. 600–607 (1998). http://www.aaai.org/Library/AAAI/1998/aaai98-085.php
6. Foxvog, D.: Cyc. In: Theory and Applications of Ontology: Computer Applications, pp. 259–278 (2010)
7. Genesereth, M.R., Fikes, R.E.: Knowledge Interchange Format, Version 3.0 Reference Manual. Interchange (Logic-92-1), pp. 1–68 (1992). http://logic.stanford.edu/kif/Hypertext/kif-manual.html

8. Hoang, D.T.A., Priebe, T., Tjoa, A.M.: Hypergraph-based multidimensional data modeling towards on-demand business analysis. In: Proceedings of the 13th International Conference on Information Integration and Web-Based Applications and Services, pp. 36–43. ACM (2011)
9. Jodłowiec, M., Krótkiewicz, M.: Semantics discovering in relational databases by pattern-based mapping to association-oriented metamodel – a biomedical case study. In: Advances in Intelligent and Soft Computing. Springer, Cham (2016)
10. Krótkiewicz, M.: Association-oriented database model – n-ary associations. Int. J. Softw. Eng. Knowl. Eng. **27**, 281 (2017)
11. Krótkiewicz, M., Wojtkiewicz, K.: An introduction to ontology based structured knowledge base system: knowledge acquisition module. Lecture Notes in Computer Science (Including Subseries Lecture Notes in Artificial Intelligence and Lecture Notes in Bioinformatics) LNAI, vol. 7802, pp. 497–506 (2013)
12. Krótkiewicz, M., Wojtkiewicz, K., Jodłowiec, M., Pokuta, W.: Semantic knowledge base: quantifiers and multiplicity in extended semantic networks module, pp. 173–187. Springer, Cham (2016)
13. Liu, J.N.K., He, Y.L., Lim, E.H.Y., Wang, X.Z.: A new method for knowledge and information management domain ontology graph model. IEEE Trans. Syst. Man Cybern. Syst. **43**(1), 115–127 (2013)
14. Pancerz, K.: Some remarks on complex information systems over ontological graphs, pp. 377–384. Springer, Cham (2014)
15. Portmann, E., Kaltenrieder, P., Pedrycz, W.: Knowledge representation through graphs. Procedia Comput. Sci. **62**, 245–248 (2015)
16. Scioni, E., Hübel, N., Blumenthal, S., Shakhimardanov, A., Klotzbücher, M., Garcia, H., Bruyninckx, H.: Hierarchical hypergraphs for knowledge-centric robot systems: a composable structural meta model and its domain specific language NPC4. J. Softw. Eng. Robot. **7**, 55–74 (2016)
17. Shadbolt, N., Hall, W., Berners-Lee, T.: The semantic web revisited. IEEE Intell. Syst. **21**, 96–101 (2006)
18. Sowa, J.F.: Conceptual graphs for a data base interface. IBM J. Res. Dev. **20**(4), 336–357 (1976)
19. Sowa, J.F.: Conceptual graphs. Found. Artif. Intell. **3**, 213–237 (2008). (Findler 1979)
20. Speer, R., Havasi, C.: Representing general relational knowledge in ConceptNet 5. In: Proceedings of the Eight International Conference on Language Resources and Evaluation (LREC 2012), pp. 3679–3686 (2012)
21. Trinkunas, J., Vasilecas, O.: A graph oriented model for ontology transformation into conceptual data model. Inf. Technol. Control **36**(1), 126–132 (2007)
22. Vandenbussche, P.Y., Atemezing, G.A., Poveda-Villalón, M., Vatant, B.: Linked open vocabularies (LOV): a gateway to reusable semantic vocabularies on the Web. Semant. Web **8**(3), 437–452 (2016). http://www.medra.org/servlet/aliasResolver?alias=iospress&doi=10.3233/SW-160213
23. Welty, C.: Ontology research. AI Mag. **24**(3), 11–12 (2003). http://www.aaai.org/ojs/index.php/aimagazine/article/view/1714/1612
24. Zhou, D., Huang, J., Schölkopf, B.: Learning with hypergraphs: clustering, classification, and embedding. In: Advances in Neural Information Processing Systems, vol. 19, pp. 1601–1608 (2007)

Application of the Nvidia CUDA Technology to Solve the System of Ordinary Differential Equations

Artur Pala(✉) and Jan Sadecki(✉)

Opole University of Technology, Institute of Computer Science, Opole, Poland
{a.pala, j.sadecki}@po.opole.pl

Abstract. This paper presents some concepts of parallel solution of ordinary differential equations in the context of the Nvidia CUDA technology. Current research leads to the development of algorithms suitable for mass parallel architecture. There has been taken into account the potential opportunities of this architecture, but also significant hardware limitations. Considered conceptions were implemented on both: the Central Processing Unit (CPU) and Graphics Processing Unit (GPU). The results of the computation time measurements and their detailed analysis were provided in this paper.

Keywords: Dynamic optimization · ODE · Nvidia CUDA

1 Introduction

This paper deals with the problem of parallel solution of systems of ordinary differential equations (ODEs). It is assumed that the nonlinear functions of the right-hand side of Eq. (1) is known. So, it is considered the initial value problem for the system of N ordinary differential Eqs. (1) [1–3].

$$\begin{aligned} \frac{dx_1(t)}{dt} &= f_1(x_1(t), x_2(t), \ldots, x_N(t), t) \\ \frac{dx_2(t)}{dt} &= f_2(x_1(t), x_2(t), \ldots, x_N(t), t) \\ &\ldots \\ \frac{dx_N(t)}{dt} &= f_N(x_1(t), x_2(t), \ldots, x_N(t), t) \end{aligned} \tag{1}$$

where:

$\boldsymbol{x} = [x_1, x_2, \ldots, x_N]^T$ - N-dimensional state vector,
$t \in \langle t_0, t_N \rangle$ - time.

The initial conditions are given as follows:

$$x_i(t_0) = x_{i0}, i = 1, 2, \ldots, N.$$

W. P. Hunek and S. Paszkiel (Eds.): BCI 2018, AISC 720, pp. 207–217, 2018.
https://doi.org/10.1007/978-3-319-75025-5_19

2 Numerical Methods

In this article, three well-known classical, sequential, numerical methods of solution of ODEs have been investigated. For these methods the influence of computational complexity on the efficiency of Nvidia's CUDA technology applications have been analyzed.

The first of the implemented methods is the Euler method. This is the simplest method from the methods outlined in this article. This method is characterized by a large integration error, which makes it also the least accurate. Its greatest advantages is however relatively great simplicity of the numerical implementation. Euler's algorithm is expressed by the formula (2) [4].

$$x_{i+1} = x_i + h \cdot f(x_i, t_i), \tag{2}$$

where:

i – iteration counter, $i = 1, 2, \ldots N$,
N – assumed number of integration steps,
h – integration step, $\mathrm{h} = \frac{(\mathrm{t_N} - \mathrm{t_0})}{\mathrm{N}}$,
$\mathrm{f(x_i, t_i)}$ – function of the right hand side of the Eq. (1).

The above description of symbols is valid for all methods presented in this paper.

The second method implemented in the developed algorithms is the Runge-Kutta II method. Many versions of this method are known. The most popular of these are the Intermediate Method, and the Huen Method. The algorithm for the Huen method can be written as follows: [4]

$$x_{i+1} = x_i + \frac{1}{2}(k_1 + k_2) \tag{3}$$

where:

$k_1 = h \cdot f(x_i, t_i)$,
$k_2 = h \cdot f(x_i + k_1, t_i + h)$.

The third implemented method, which is the most complex of the mentioned ones, is the Runge-Kutta IV method. This is the most well-known and most widely used method from the wide class of the Runge-Kutta algorithms. This method provides the relatively good accuracy of computations at relatively large value of the integration steps. The disadvantage of this method is that it is necessary to calculate the value of the function on the right hand side of the differential equation four times in each step of integration. In addition, these values are no longer used in any further computations. The algorithm of this method is expressed by the formula (4) [4].

$$x_{i+1} = x_i + \frac{1}{6}(k_1 + 2k_2 + 2k_3 + k_4) \tag{4}$$

where:

$$k_1 = h \cdot f(x_i, t_i),$$
$$k_2 = h \cdot f(x_i + k_1/2, t_i + h/2),$$
$$k_3 = h \cdot f(x_i + k_2/2, t_i + h/2),$$
$$k_4 = h \cdot f(x_i + k_3, t_i + h).$$

3 The Concepts of the Parallel Solution of Ordinary Differential Equations in the Context of Nvidia CUDA Technology

Solving of the large systems of ordinary differential equations (ODEs) is still a challenge for the latest multi-core CPUs, whose architecture has built-in capability of parallel processing, but still on a relatively small scale. A completely different kind of architecture and kind of computational organization is provided by Nvidia CUDA technology used by GPUs. These differences in comparison to the CPU structure creates a very efficient parallel processing model on a mass scale. The great problem is that still there no global computational methods and algorithms capable of harnessing the potential offered by Nvidia's CUDA technology. In addition, the currently used numerical methods of solution of ODE have the strictly sequential nature, which makes it impossible to implement them directly as the parallel algorithm, especially in the Nvidia's CUDA architecture.

In references [5–7] it have been shown concepts of parallel solution of ODE systems. Some of these concepts offer a sufficiently high degree of computation parallelism which make possibilities to use Nvidia CUDA technology. They are based on the use of known sequential methods. One of the such idea assumes division of the integration interval into a number of sub-intervals that are distinct problems, so each of which can be solved independently by one of the known sequential methods. However, this requires estimating of the values of the elements of state vector at the begin of each sub-interval, before the proper integration of the system of equations. The final solution is generated by the choice from the all sub-interval and all groups of solution, which is characterized by the most closely matched values of solution in the points between sub-intervals. This idea assumes that there exist not one, but many of the initial conditions for each element of the state vector. So, consequently the set of initial conditions is created as it is shown on Fig. 1.

Computations in each of the sub-interval runs for all generated initial conditions, creating multiple alternative groups of solutions. The final solution is constructed by the select from each sub-interval and from each group of solutions most closely matched values between sub-intervals.

This idea named as the Speculative Method was presented in the PhD thesis [8]. The author has made the computations on the base of example of analysis of transient states for some electrical systems. For the solution of system of ordinary differential equations the Runge-Kutta IV method has been used. As the computing environment a cluster system dedicated to conducting scientific calculations (The Department of

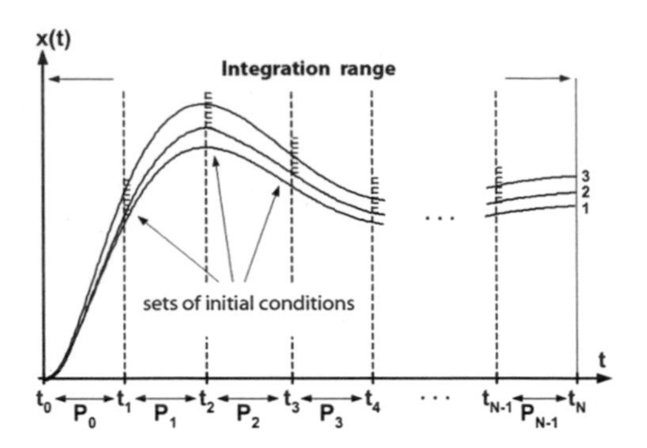

Fig. 1. Scheme of divide problem into N of subproblems [8]

Theoretical Electrical Engineering and Metrology of Bialystok University of Technology) has been used. According to the author, the cluster was built with 7 nodes working under Linux, and communication between them was implemented using the MPI standard. The aim of the study was the development of a parallel numerical method of analyzing transient states of electrical circuits, which allow to cut the time of transient analysis. The maximum acceleration of the computations achieved by the author in all of many analysed cases was 4.09.

It must be noted that the presented concept has greater potential in parallel calculations than presented in the paper [8]. The parallelism offered by this concept is limited only by the number of available processors, which do not need to communicate with each other. These features qualify this concept to research into the possibilities of its implementation in Nvidia CUDA technology.

An alternative idea is to decompose the system of differential equations into blocks of few equations (or into few blocks of single equations) that can be computed in parallel (Fig. 2).

$$\mathrm{Processor}\ \ 1 \leftarrow \dot{x}_1 = f_1(x_1, x_2, \ldots, x_N, t)$$

$$\mathrm{Processor}\ \ 2 \leftarrow \dot{x}_2 = f_2(x_1, x_2, \ldots, x_N, t)$$

$$\ldots$$

$$\mathrm{Processor}\ \ N \leftarrow \dot{x}_N = f_N(x_1, x_2, \ldots, x_N, t)$$

Fig. 2. Distribution of system equations between processors/CUDA cores

This idea does not give such great possibilities of parallel calculations as the Speculative Method because it is limited by the number of differential equations of the system. In addition, the concept assumes synchronization and communication between threads at each integration step, which further limits the possibility of division of work between two separate CUDA devices. However, it assumes a lower computational cost and may be more efficient than the Speculative Method when the number of equations in the system oscillates around, or even exceeds the number of available execution cores in the GPU. Any concept of implementation of parallelism at the level of the numerical method (Eqs. 2, 3, 4) was rejected due to the low possibility of implementation of parallel computations on this level of algorithms, which do not allow to fully exploit the potential of Nvidia CUDA technology in this case.

4 Parallel Algorithms for Solving of Systems of ODEs

Three ODE algorithms for Nvidia CUDA technology have been developed. Because of their properties and different organization of computations, these algorithms are a mutual alternative and are dedicated to different areas of application.

The first of the developed algorithms uses the assumptions of a concept that divides the system of equations into individual equations or groups of equations. This concept will be referred as the concept of parallel equations. The work of the algorithm can be divided into three main stages. The first stage is implemented on the Host side and involves, among other things, allocation of Host memory, loading of data and placing them in appropriate memory structures, allocation of CUDA memory, copying data to device memory and division of the system of equations into few group of equation or few single equation, depending on their number in the comparison to the number of available cores.

The second stage is implemented on the CUDA side and involves parallel execution of calculations, synchronization and communication operations between individual computing threads at each step of integration. Each thread executes calculations using one of the known sequential numerical methods. The third step is done on the Host side and includes copying the results from the CUDA memory to the Host memory and releasing CUDA resources.

The second proposed algorithm uses the concept based on the divide of the integration interval into a number sub-intervals. This idea generates additional computational problems. This concept will be called as the Speculative concept. In this case the work of the algorithm can be divided into three main stages. It should be noted that in this algorithm, the first stage is much more complex and computationally complex than as it was in the case of the first algorithm. The first stage of the algorithm works on the Host side and includes such tasks as, allocation of Host memory and loading of data into it. The integration interval will be divided into P of subintervals. Next, the differential equation should by twice solved: first for the $h = (t_N - t_0)/P$, and next for the $h = 0.5h$. On the base of these two solutions, the initial condition for each of subinterval can be designated [8].

Based on the difference in equation solutions obtained for both steps, at the beginning of each integration sub-interval a predetermined number of initial conditions (Nv) is generated. On the CUDA device the allocation of memory is performed, and data are copied from Host to the device. The second stage is realized by the CUDA device, where the system of (Nv) of speculative equations should be solved for each of the independent integration sub-interval (P). For the subintervals, computations are performed using the selected sequential numerical method. The third and final stage is the copying of all speculative partial solutions into Host memory. Later, the Host selects from the all solutions the solutions which fit most closely together in the points between each pair of adjacent subintervals (Fig. 3).

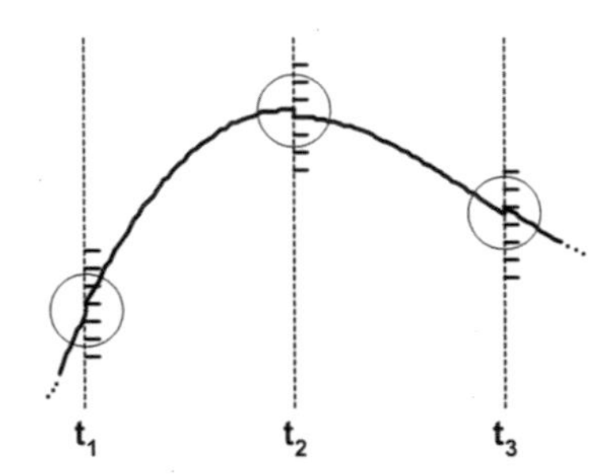

Fig. 3. Final solution created by partial speculative solutions [8]

The third algorithm is a variation of the second algorithm. The difference between them lies in the different implementation of the second stage. It has been noted that in the case where more than one equation is solving, we can obtain a greater degree of computation parallelism by the parallel implementation of solving of equations within each of the integration subintervals. The third algorithm is based on two concepts, the concept of parallel equations and the concept of speculation.

5 Hardware Limitations of Nvidia CUDA Technology

The first major limitation of CUDA technology, based on the effectiveness of the developed algorithms, is that the CUDA device is a separate subsystem with its own resources in the computer system. This generates a series of additional actions that Host must perform to run calculations on a CUDA device. Especially in some cases copying of data to the device and receiving calculation results is very expensive. In addition, for the first of the developed algorithms, based on the concept of parallel equations, the possibility of working on many separate CUDA devices is very limited. In this case, the communication between separate devices during each integration step unfortunately must be run by the Host.

The second limitation is communication between threads. For the first of the developed algorithms, the groups of equations correspond to a group of threads, called blocks. All available GPU cores are also split into groups, called multiprocessors. Individual thread blocks work on separate multi-processors, and communication between them is limited, and must take place via the global memory of the CUDA device. Global memory is, however, characterized by relatively long data access times. In addition as to global memory, CUDA device also has a very fast shared memory that is physically located inside the GPU. However, this memory is available only for threads working in the same block, so within one multiprocessor. In addition, access to individual shared memory banks should be based on special rules, because conflicts between threads [9] are possible. Due to the assumptions of the parallel equations concept, synchronization and task communication between computing threads occur multiple times in each integration step, generating multiple access conflicts. A typical conflict occurs when threads duplicate their access requests to variable values computed by other threads (Fig. 4). This largely limits the use of shared memory.

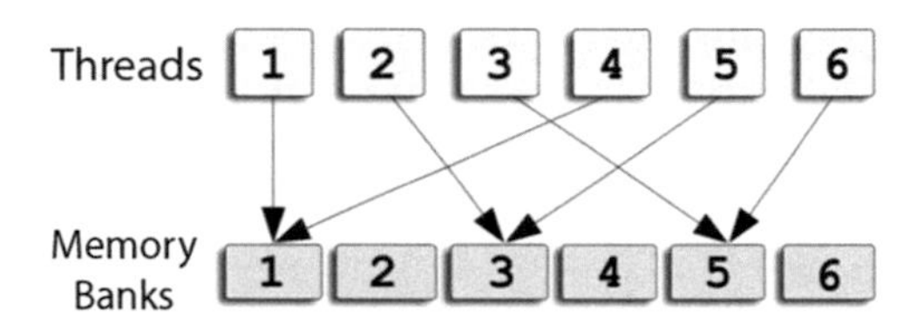

Fig. 4. Conflicts between threads [9]

In spite of the limitation, the algorithm uses part of shared memory by storing corresponding rows of matrix coefficients in it. This limits the number of references to the global memory of a CUDA device when solving the equation in subsequent integration steps.

The restriction does not apply to the second of the developed algorithms, which is based on the Speculative concept. This is because there is no need to communicate between threads working on different multiprocessors or between threads working within a single multiprocessor, since each thread resolves a separate speculative set of equations in a separate integration sub-range.

The limitation of communication between threads is, however, related to a third algorithm, then based on both concepts. It is less than the first one because there is no need communication between threads working on separate multiprocessors, and only communication within the multiprocessor is required.

6 Time Measurement for Algorithms

For the simplicity only linear model of Eq. (1) was considered in the paper. In this system, defined: $\dot{\boldsymbol{x}} = \boldsymbol{A}\boldsymbol{x}$, elements of matrix $\boldsymbol{A}$ and initial conditions were generated randomly. The considered systems of equations were solved by use of all version of developed algorithms. All of those systems were solved both using sequential numerical methods on the CPU and using parallel algorithms on the CUDA device. The software also implemented a detailed measurement of the working time of all developed algorithms.

The measurement platform for an algorithm based on the concept of parallel equations was a computer: CPU: AMD 64 X3 3.3 GHz, 4 GB of RAM, CUDA Device: GeForce 650 GTX Ti, Windows 7 operating system (64 bit version).

For the algorithm based on the Speculative concept, the measuring platform was a computer: CPU: 2 x Intel Xeon E5-2620 v2 2.10 GHz, operating memory: 32 GB, CUDA device: GeForce GTX TITAN, Windows Server 2012 operating system.

7 Results of Computations

Within the concept of parallel equations, the working time of the sequential algorithms is indicated by the letter S. Parallel algorithms are marked with the letter R. Parallel algorithms using fast shared memory are labeled with RSM (Table 1). The results of the measurements are presented in three cases, each based on the use of numerical methods of Euler, Runge-Kutta II, and Runge-Kutta IV.

All time values obtained from measurements are given in milliseconds. The exact time values of all presented measurements are shown in Table 1.

Table 1. Time measurement results in milliseconds

Case	Algorithm	Number of threads/equations	Sequential numerical method		
			Euler [ms]	Runge-Kutta II [ms]	Runge-Kutta IV [ms]
1	S	1/120	111,996	219,104	425,268
	R	120/120	208,041	272,026	391,762
	RSM	120/120	101,938	114,935	152,755
2	S	1/300	608,769	1189,509	2356,308
	R	300/300	450,227	627,876	965,058
	RSM	300/300	183,231	215,200	311,949
3	S	1/50	32,227	51,996	86,111
	R	50/50	90,130	113,960	138,729
	RSM	50/50	75,047	79,087	88,192

The number of integration steps for all cases was 1000. There were also measurements with a different number of integration steps for all cases. However, this did not significantly affect the proportions between the working times of particular algorithms. The parallel algorithm (R) gains an advantage over sequential implementation (S) only for the Rungego-Kutta IV method. However, the algorithm using shared memory (RSM) was faster than sequential (S) for all numerical methods. In addition, the Runge-Kutta IV (RSM) method was faster than (R) by approximately 61%. In case 2 (Fig. 5), parallel algorithm (R) worked faster than sequential (S) for each of the applied numerical methods. In addition, the Runge-Kutta IV (R) method was 41% faster than (S), and (RSM) was 68% faster than (R). The third case presents a negative situation from the point of view of the efficiency of the parallel algorithm. For all numerical methods, algorithms (R) and (RSM) worked longer than sequential (S).

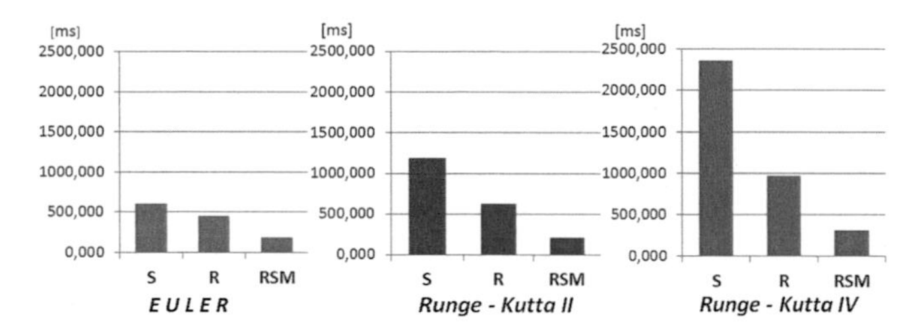

Fig. 5. Time results for the second case

The results of working times for developed algorithms based on the Speculative concept are presented in Table 2, in terms of three cases, each with a different number of solved equations. The sequential algorithm's working time is indicated by the letter S, for the algorithm based on the speculative concept, the SP sign is assumed. On the other hand, the algorithm based on concepts of parallel equations and the Speculative concept was designated SP+RR. For this group of algorithms it is important that the maximum achievable accuracy of the solution stands at the level of the sequential solution. This limitation is due to the assumptions of the Speculative concept. The accuracy of the solution depends on two parameters, the number of sub-ranges (*P*) and the number of speculations (*Nv*). Studies have shown that the quickest improvement in accuracy is achieved by increasing the number of integration sub-ranges (*P*). However, there is some limitation due to the fact that by increasing the number of sub-ranges (*P*) we also increase the amount of work that Host must perform in the first stage of the algorithm.

Table 2. The time measurement results in milliseconds

Case	Algorithm	Number of threads/equations	P/Nv	Sequential numerical method			
				Euler		Runge-Kutta IV	
				Time[ms]	Norm	Time[ms]	Norm
1	S	1/1	–	24,6571	–	131,7622	–
	SP	400/1	100/4	49,4523	0,8303	82,8975	0,0276
	SP+RR	400/1	100/4	50,2365	0,8303	89,4399	0,0276
2	S	1/3	–	114,8853	–	483,8622	–
	SP	400/3	100/4	177,0992	1,0308	503,6486	0,5950
	SP+RR	1200/3	100/4	108,3928	1,0308	210,4013	0,5950
3	S	1/6	–	348,0382	–	1654,6711	–
	SP	400/6	100/4	483,5720	10,7978	2508,6630	7,5444
	SP+RR	2400/6	100/4	219,8555	10,7978	374,8135	7,5444

There is a P value for which the use of the parallel algorithm is no longer time-efficient with respect to the sequential solution. For all of the cases presented in Table 2, the number of integration steps was set at 1 000 000, the P value was empirically determined at 100, while the number of speculations (*Nv*) to 4.

The third aspect affecting the accuracy of the solution is of course the accuracy of the applied sequential method. For all analysed cases, the SP and SP+RR algorithms for the Runge-Kutta IV method are much closer to the sequential solution than it is in the case of the Euler method. This illustrates the parameter β, calculated as the Euclidean norm (this parameter is denoted in the Table 2 as Norm):

$$\beta = \left(\sum_{i=0}^{N} (y_{1i} - y_{2i})^2 \right)^{0,5},$$

where β is the value of the norm, *N* is the number of integration steps, *i* is the index of the elements of the solution vector, *y1i* and *y2i* are the i-th value of the solutions of the sequential and parallel algorithm respectively.

It should be noted that the increase of the number of solved equations negatively affects the accuracy of the parallel solution, which moves away from the sequential solution. Also, in analyzing the results from Table 2, it should be noted that in case 2 and 3, the SP+RR algorithm allowed for a much greater parallelism of calculations than the SP algorithm with the same values of parameters *P* and *Nv*. This algorithm is much faster not only for the algorithm SP but also for the algorithm S. In the first case, for the Runge-Kutta IV method, the SP algorithm was the fastest.

8 Conclusions

The results presented in this paper clearly show that each of the developed algorithms is effective in slightly different areas of application. In addition, the algorithm based on the Speculative concept can provide unexpected results when applied to equations with oscillating nature. In this case, we would probably need to create much more initial conditions at the beginning of each of the integration sub-ranges. In this situation, there is a risk that the algorithm based on the Speculative concept will prove too costly and unprofitable compared to the classical sequential algorithms. However, this problem has not yet been investigated by the authors of this paper. It will be the subject of further research. The fact that the algorithms complement each other makes it possible to use Nvidia CUDA technology to solve a relatively wide spectrum of systems of ordinary differential equations. To confirm this thesis we are working on creation of new dedicated algorithms for CUDA technology which are based on wide class of algorithms for solving of systems of ordinary differential equations. These methods can be also used for instance for solving of dynamic optimization problems which a based on multiple solution of large systems of state equations [10], so they fit well into the applications of the developed algorithms and Nvidia CUDA technology.

References

1. Baron, B., Kolańska-Płuska, J.: Numerical Methods of Solving of Differential Equations in C#. Opole University of Technology Press, Opole (2015). (in polish)
2. Niedoba, J., Niedoba, W.: Ordinary and Partial Differential Equations. AGH University of Science and Technology Press, Krakow (2001). (in polish)
3. Gewert M., Skoczylas Z.: Ordinary Differential Equations. Theory, Examples, Tasks. Publishing House GiS (2005). (in polish)
4. Hapra, S.C., Canale, R.P.: Numerical Methods for Engineers. The McGraw-Hill Companies, Delhi (2011)
5. Burrage, K.: Parallel and Sequential Methods for Ordinary Differential Equations. Clarendon Press, Oxford (1995)
6. Gear, C.W.: Massive parallelism across space in ODEs. Appl. Numer. Math. **11**, 27–43 (1993)
7. Petcu D.: Parallelism in Solving Ordinary Differential Equations. Mathematical Monographs 64, Timisoara University Press, Timisoara (1998)
8. Forenc J.: Speculative analysis of transient states in electrical systems, Ph.d. thesis, Bialystok University of Technology (2006). (in polish)
9. NVIDIA CORPORATION: Nvidia CUDA Programming Guide, Nvidia Corporation (2012)
10. Findeisen, W.ł., Szymanowski, J., Wierzbicki, A.: Computational Methods of Optimization. Warsaw University of Technology Press, Warsaw (1973). (in polish)

Towards Semantic Knowledge Base Definition

Marek Krótkiewicz[1], Krystian Wojtkiewicz[1,3(✉)], and Marcin Jodłowiec[2,3]

[1] Department of Information Systems,
Wroclaw University of Science and Technology,
Wrocław, Poland
{m.krotkiewicz,k.wojtkiewicz}@kieg.org
[2] Institute of Computer Science,
Opole University of Technology, Opole, Poland
m.jodlowiec@kieg.org
[3] Institute of Control Enigneering,
Opole University of Technology, Opole, Poland

Abstract. The paper is a wide survey over one of the knowledge representation and processing solutions, namely knowledge bases. Due to current terminological inconsistency authors propose the complex definition of knowledge base in the field of knowledge representation. The overview of the most common reality description methods is provided in order to discuss its usefulness in knowledge base design. Authors not only give the definition of the knowledge base but also prove its completeness on the example of Semantic Knowledge Base project. The project aims at developing the general domain knowledge base using ontology base and semantic networks as basic knowledge representation methods.

Keywords: Knowledge representation · Knowledge base · Ontology
Semantic network

1 Introduction

Knowledge has been the source and target of studies from the very beginning. Plato was probably the first philosopher who tried to set the definition of knowledge. He started from "Knowledge as perception", to propose another "Knowledge is True Judgment" to finally provide definition, that is: "Knowledge is true judgment with an account". The Plato's Theaetetus [3] just started the discussion that still lasts. After over 2000 years, we still cannot provide one universal definition of Knowledge. However for the purpose of this article we will define it as "Data with information how to use/interpret it".

Knowledge design and processing is one of the most complex contemporary research issues in the computer science. The design of the knowledge base is crucial for the effectiveness of the expert system build over it [10]. Main problem associated with this task is the availability of models flexible enough to build not only data structures but also structures for storage of knowledge. Relational

W. P. Hunek and S. Paszkiel (Eds.): BCI 2018, AISC 720, pp. 218–239, 2018.
https://doi.org/10.1007/978-3-319-75025-5_20

model, that is the most common model in databases design, is at the same time very inefficient in the means of complex, hierarchical data structures [6,13,27,31].

The description of complexity of the reality is an important aspect of research regarding artificial intelligence. The study developed many different approaches that not only intermingled with each other but most importantly they enabled to increase the expression of the phenomena and facts described. Below, the methods of describing the reality conceptualization most frequently mentioned in the literature of the subject are presented.

2 Basic Methods of Conceptualization of Reality

2.1 Abstraction

One of the basic methods used from the beginning of modelling. Its main assumption is the principle of functional decomposition that is based on defining operations as more elementary operations. Each level of decomposition leads to the increase of detail in the description of primary operation by introducing new elements of description that on the higher level of abstraction are not important. As the example, let us consider the operation of "building of a house" as the highest level of abstraction. Defining such an operation the following sub-operations can be specified:

1. House construction preparations,
2. House construction realization,
3. House construction finalization.

The presented lower level operations are only the examples. By creating the hierarchic structure of the operation the aspects that may constitute the independent dimension of this description can be taken into account. For example, the operation "house construction preparation" can have its aspect (dimension), e.g.: financial, organizational, technical.

In each of this dimensions there is an independent hierarchy of the operation in the sense of division into aspects, what was presented on the Fig. 1, what is not contradictory to the possible relations between operations only, in the functional sense.

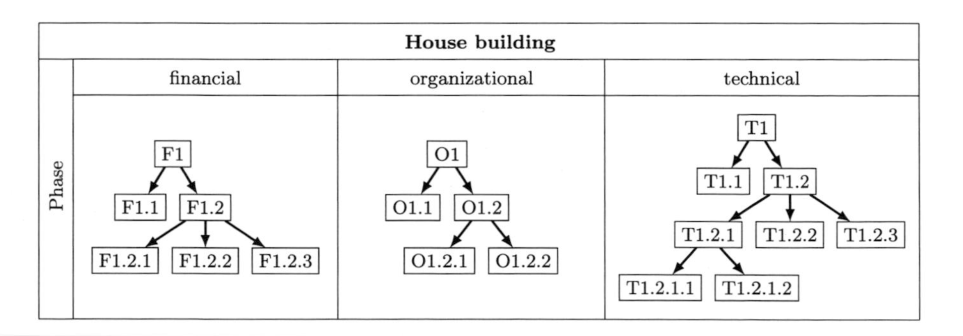

Fig. 1. House building abstraction

For instance, the preparation in the technical and organizational sense will certainly depend on the financial aspect. Moreover, there are other close relations between organizational and technical aspects. In practice, it means that procedural abstraction can be transformed from the simple tree to the form of multidimensional labelled directed graph, defined as:

$$G = (V, E, \gamma) \tag{1}$$

where:
V – set of nodes,
E – set of edges,
γ – function over nodes and edges.

In the classical approach to the procedural abstraction [20,28] this issue comes down to the principle, according to which any operation that reaches defined goal can be treated as a whole, regardless to the fact that the operation itself may actually consist of a sequence of lower level operations.

The abstraction can also be used to describe not only operation but also data. Data abstraction is based on their description in the operation categories. This idea is very strongly used in the object-oriented programming paradigm, where, in contrast to the procedural paradigm, a strong connection of structures (classes) with operations (methods) that can be performed on the instances of this structures (objects) has occurred.

2.2 Encapsulation

It is the concept that is based on the principle of minimization of information available outside. An element makes the information available through interfaces, that is set methods of communication [12,25,30]. This method of reality description ensures that the surrounding "knows" only what is absolutely necessary and subsequently there is no access to inner information of encapsulated element. It is very important from the point of view of consistency, because it protects encapsulated element from the unauthorised access that can involve reading or modifying the value of this element. In object-oriented encapsulated modelling the element is a class, where by default all components have private visibility range. As a result, all that has not been explicitly declared as public may be modified only by the operations defined in this class. This protects items of a particular class from the modification of attributes or from performing an operation that is reserved only for the inner purposes of items of a particular class. Thus, the encapsulated element has private (hidden) and public (accessible from the outside) part, whereby the principle of encapsulation implies that only those elements that need to be public are public, in contrast to others, in particular procedural approach, where there has been no mechanism protecting the structure of data and functions against misuse.

From the point of view of artificial intelligence the example of encapsulation may be the abstract model of reality called the Chinese Room [8]. In this approach encapsulation is based on hiding grammatical rules and communicational language semantics, in contrast to the object-oriented model, where data and operations are considered.

2.3 Inheritance

It is the concept of reality description focused on the relation between generalization and specialization. This description comes down to the creation of taxonomic trees (Fig. 2), in which the superior elements (base elements) describe the common set of features, properties, operations for all derivative elements (inherited elements). This mechanism is highly important and commonly used to the description of phenomena, both in information technology and in other fields. There are many relations between this and other mechanisms. For instance, in object-oriented modeling the idea of inheritance is closely related to such issues as encapsulation, visibility ranges and interfaces, resulting in coherent description (modeling) system that has a wide range of expression.

Inheritance has several aspects. One of them is the problem of multiple inheritance that is based on the possibility of inheriting from more than one base element. The result is that the classical generalization tree becomes a directed graph (acyclic graph).

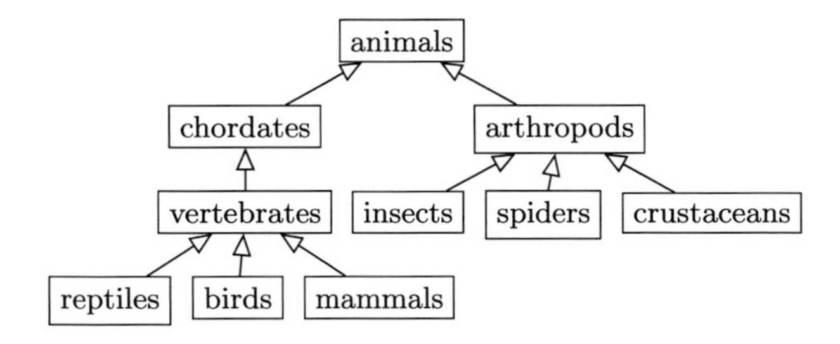

Fig. 2. Exemplary animals taxonomy tree

Multiple inheritance is a strong description tool, because it enables the simple and quite intuitive method of presenting a situation, in which a particular element constitutes a kind of hybrid of two or more superior elements, regardless of whether it refers to structures, operations or other elements that are subject to inheritance (Fig. 3).

The issue of inheritance involves the problem of virtuality that protects against unauthorised duplication of elements or of different classes of components that are in fact the same element. It is particularly important in the inheritance structures called diamonds or their derivatives (Fig. 4).

2.4 Connotation

The description of reality through the associations is based on creating relations between elements that in some aspect or dimension have common or similar features, properties or functionalities [2,7]. Such relationships are ambiguous or unnamed and they do not specify precisely the nature of this relation but rather loose connections between elements. It is also restricted to the specified aspect.

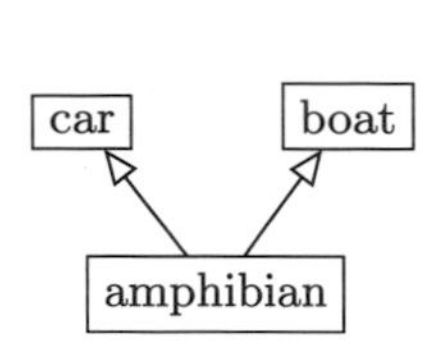

Fig. 3. An example of multiple inheritance

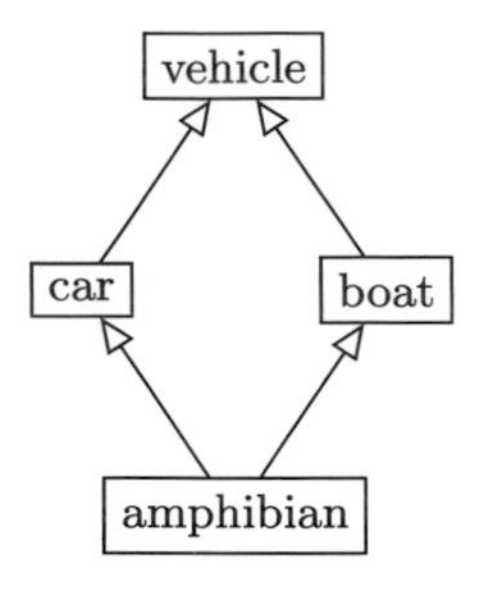

Fig. 4. An example of diamond structure in a inheritance graph

For instance, some events may be associated in the dimension of time, what would suggest that they can occur in e.g. in the close distance of time from each other, they can constitute consequents or precedents to some other events. As a result, associations are inherent to the concept of similarity, however, they do not imply the necessity of a strict definition of metrics but only to more or less set out the proximity in a particular dimension or space. Other more complex aspect of similarity that constitutes the basis for associations is for instance the structure. The structure, as the system composed of elements and relations between them. The classic example of associations related to the aspect of structure is the term "tree" that is associated with many concepts including the concept of a physical vegetal object and the abstract concept of data structure that derives from the graphs theory. Completely independently, but close to that concept is the concept of e.g. decision tree used in expert systems of artificial intelligence. Analogically, we can list directory trees, trees as organizational structures in management sciences, taxonomic trees, trees defining inheritance etc. In this case, the combining (associating) elements are the features of structure referred to as the tree, which here is an incoherent directed acyclic graph.

2.5 Scale

The method of description based on the scale relates on the idea to look at the described model from a distance that causes some details to be invisible [29]. This idea is quite natural, however it is highly important not only for the description of reality in the abstract sense, e.g. knowledge base, but also it constitutes a good illustration for the method known in the modelling of 3D scenes named "elision". It is based on optimization of number of displayed polygons (the smallest element of which 3D scene is build). This optimization is focused on the reduction of polygons, that is details of displayed items, in the function of distance of element and observer. We can imagine a very complex scene composed of a very large number of complicated solids that additionally change their location in time creating a dynamic scene. The example of such scene can be a dynamic 3D image presenting an epic battle, which involves 10 000 warriors, each composed of thousands elements that are polygons. If the observer is able to see the whole scene, it raises the question if it would be wise to overload a processor by the necessity of calculation the location and other parameters for each of this billiards of polygons. Taking into account that by transforming this 3D scene to the form of projection onto a plane, individual warriors will cover only a few pixels. Thus, the number of processing elements of a 3D scene depends on the distance from observer, so the processor is not overloaded with the necessity perform calculations on details that will not even have a chance to appear in the form of visual effect.

And this is a key to the idea of reality description based on the scale, where by presenting any model, some of its details needs to be hidden and which of them will be hidden, and specifically to which level of significance the element is visible, depends on the scale at which the model is presented. It has extremely significant for the ability to perceive the model and its effective analysis. Entering into details on every level of zoom by the observer will cause the necessity to take into account too many details. This is why, there are several levels of model granularity. From the detailed model the elements can be segregated so that the model starts to be “light” and easy to analyze and comprehend by a person or persons working on it.

The idea of scale is commonly used for a long time during the use of all kinds of maps. There are general maps, where placing of details would obscure the image, if at all it was possible because of their richness. However, the detailed maps can and should consist of possibly a large amount of information, so that the change of scale is not a simple operation based on mathematical homothetic transformation, and at the same time provide new information hidden on the higher levels of scale [14]. The scale can relate to many aspects, e.g. it can be connected with the relation between a part and a whole. For example a technical drawing, where some of the parts can be visible and some of them hidden and what decides about it is the scale of such drawing.

It can also relate to more abstract relations, e.g. generalization-specialization, where while considering for instance taxonomy, we can focus on the upper part of a tree going down to the direction of leaves that are in fact elements that in a specific scale are seen as the last, the darkest and the most detailed. In this case the scale means moving inside the tree of taxonomy.

3 Complex Methods of Conceptualization

3.1 Object-Oriented Modeling

Object-oriented modelling [21] is an example of idea, in which several methods of modelling were combined, including a few presented above. Therefore, the description of features of such modelling focuses only on its selected elements, in order to avoid redundancy of description. One of the more important concepts of object-oriented modelling is class-object relation. The class is a description of a construction and the way of creating objects, whereas the object itself constitutes the “physical” emanation of class, as its instance. It is necessary to emphasize that the object is not a part of the class and it is able to store information in attributes and to interact with other objects through methods. In this approach methods are part of the class, that is they are defined in it, whereas they are induced for objects. In a specific case, the object can have an interaction through the method with itself; then it refers to the methods induced for and inducing object. In the natural language it is the equivalent of inducing operation described as the reflexive verb. (an example: washing yourself) Another already mentioned concept are the attributes. Attributes are defined in the class but the ability to assign them the values is reserved for the objects. Due to that, all objects of a particular class have the same set of attributes but each of the objects can consist of other set of values assigned to these attributes. A special case of attribute is static attribute, where it is defined and stored on the level of the class and all objects of a particular class referring to this attribute, refer to one and the same element. Object-oriented modelling does not imply aggregates that have the ability to store objects, that is in the same idea there is no separate category with such function. In practice, it means that each implementation can, according to its rules, create sets, lists, vectors, bags and multisets, however, as such, they are not the element of object-oriented modelling.

One of the basic method of object-oriented modelling is the concept of association. Association is the connection that links two classes. This relation is unnamed, which means that its semantic is not predetermined, in contrast to e.g. composition or generalization. Association has some other features, such as navigation, multiplicity of association or roles, however these aspect will not be discussed in detail in this article.

The object-oriented model despite its size and generality has restrictions and does not include all possible concepts of description of world complexity.

Considering the restrictions, we should notice that in many cases they rather do not result from the model itself but from its implementation. For instance, already mentioned association normally is realized by the attributes. The result is that the relation of association automatically became purely conceptual, that is it does not consist of its own separate categories or their instances. Simply put, in languages using object-oriented paradigm usually it is impossible to create an entity that would be an instance of association. The special case of association are n-ary association, for $n \neq 2$. Each of the roles of this association consist of multiplicity, but only on the side of a class. As a result, it is possible to limit the number of objects that take part in association, but it is impossible to determine the number of association, in which the object of a particular case can take a part. This stand in a sharp contrast to the binary association, where on both sides there are multiplicities, which means that ternary associations and higher are a special case that should be consider separately in relation to the binary associations. It is inconsistency that goes much further, because in the model unary relations were not included, although they are useful in the modelling.

By reviewing the existing methods of the world complexity and researching their mutual relations, way of mutual complementation, redundancy, strengths and weaknesses, what was developed is the skeletal system of a knowledgebase which in the possibly most general way is able to store the information of the possibly most general character. This system is hybrid both from the point of view the conception of complexity description and in respect of the character of data that are stored there. It connects many simple and complex ideas, the latter of which are often extended versions of their primary concept.

3.2 Ontology

Ontology in the sense of science is the area known to philosophy and it is defined as "as it is". It means that ontology consists of information about the state of things (state of the world), entities and relations between them. From the point of view of information technology, it can be determined that ontology is focused on entities and relations between them, not, however, on the answer to questions why is it like this or how does it happen, etc. The most popular definition of ontology in the information technology is Tom Gruber's phrase saying that ontology is "*specification of conceptualization*" [9]. This definition is elegant because of its simplicity and generality, however it does not explain too much. From the moment, when the concept of ontology in information technology became commercialized, it started to be identified with the particular solution, and even with implementation of this solution in the form of OWL[1] and its derivatives. Ontology in this sense has lost on its generality and moved away from that idea, indicated in 1994 by Tom Gruber. Moreover, many studies present the specific implementations based on e.g. taxonomy, pompously called ontology of a particular problem. Sometimes it is even called ontological base, which is a terminological mistake, because ontology itself constitutes a structured set of information having a specific structure [5].

Aside from the issues of terminology, ontology is undoubtedly one of the most important methods of world complexity description. It should, however, be emphasised that it surely does not exhaust all the aspects of this description. In particular, ontology is not intended for storing the rules and facts beyond the entities and relations between them. Ontologies are usually presented as sets of types (classes), objects (instances) and relations (associations), which can be connections of any type. In practice, solutions based on ontologies very often make a predefined assumption, that is predetermined number of types (classes) and kinds of relations. Restriction that is particularly strong is the determination of kinds of relations, usually for generalization-specialization, part-whole and other for special uses. Lack of possibility to freely define relations is particularly burdensome for a modelling person and it significantly narrows the possibilities of a specific system.

3.3 Semantic Networks

Semantic networks [4] are the method of complexity description based of the graph theory. In semantic networks there are nods and edges, whereas both nods

[1] Web Ontology Language – a family of knowledge representation languages endorsed by the World Wide Web Consortium (W3C).

and edges consist of labels (Fig. 5). What is particularly important for the semantic networks, is the fact that they do not have predetermined semantics. It is their fault, but also a huge advantage. It can be a fault because when seeing any defined network without the description of its semantics, it is impossible to clearly determine the meaning of nods and arches. It means that by creating any semantic network, it is necessary to precisely determine the semantics of arches and nodes and possible grammatical structure accepted by these networks. It is at the same time an advantage, because it proves the generality of this method and allows to use them for the description of incredibly wide range of problems: from simple arithmetical equations, which can be easily represented by such networks, to complicated rules or facts that apply to the artificial intelligence systems. The simplicity of a structure of such network, that is nodes, edges, their labels and a function determining which of the nods are connected with which and with which arches, allows to express the complexity of the world. However, the classical semantic networks also have some disabilities [17].

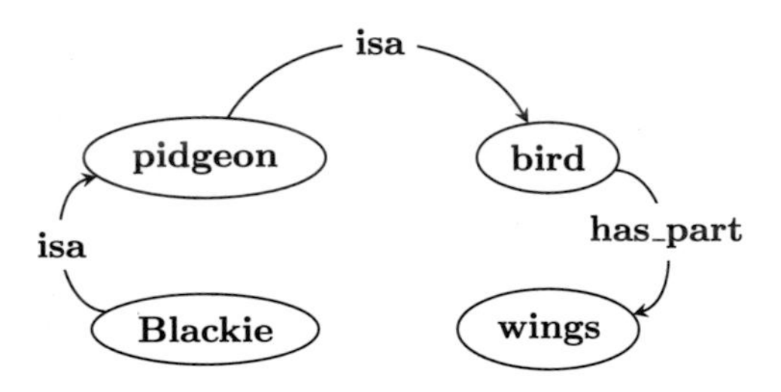

Fig. 5. An exemplary semantic network

They are, for instance, unable to take into account some of the details important for the description of reality. Many problems relating to the semantic networks have been solved in the Hendrix's studies over 50 years ago [11]. Unfortunately, the literature review proves that incredibly small number of authors describing the networks include the solutions proposed in this studies, referred to as the partitioned networks. The mechanism described by Hendrix is strong and it ensures the hierarchic structure of a network. Still, there are some issues left that in the classical networks and in networks extended by Hendrix are not included. An example of which is a mechanism of associations multiplicity, which is not present in the classical semantic networks. The problem relating to associations multiplicity id practically insolvable on the level of graphs and only after introducing the structures known as hypergraphs[2] to the description of semantic network structure, the problem is possible to be modelled. Semantic networks in the classical approach consist of more important restrictions, which, however, are not the subject of interest of this article.

[2] Hypergraph's edges are called hyperedges; they can be incident to any number of vertices.

3.4 Natural and Artificial Languages

The most natural way of complexity description is the natural language [22–24]. Its undeniable feature is its incredible flexibility allowing to describe almost every idea. It is, however, burden with a very difficult and, surprisingly unnatural for a human, method of recording. It is one dimensional recording that is linear and strongly marked with cultural connotations and connected with the set of terms and concepts being at the authors disposal for a specific statement in the natural language. It is a huge obstacle to the analysis of a text, because the only way for its acquisition is to combine terms as sequences that needs to be in accordance with a set grammar of a natural language. Already mentioned unidimensionality excludes the possibility to move freely on the record. The reality described is usually multidimensional and multifaceted and can include many levels, points of view and transformation of very complex information to the form of a sequence of terms. This needs to cause, first of all, difficulties, second of all, illegibility, and the third, a huge effort of the receiver to transform this sequence back to the complex structures that are functional for our minds. This is why, the natural language has been displaced from the technical sciences, especially from information technology, where two-or three-dimensional structures are dominant, also constituting the (often graphic) representation of a concept, idea or basically the complexity of the world. An example for this could be dozens of diagram types used for the description of equivalent issues. What is used within the modeling of information systems is for instance UML, which alone consists of several number of diagrams. Additionally, there are many diagrams in other sciences, such as BPMN [1], which is widely used in the combination of information and management science.

Formal languages [18] have been created for a very specific needs. Currently, the primary most universal formal language is language of logic, on which other fields of mathematics were based. Formal languages, e.g. programming languages, consist of very strict grammar and semantics, what differs them from the natural language. They are so unambiguous that it is possible to design machines able to communicate in such way. It is, however, burdened with illegibility for an average person and often with a considerable complexity of records. It should be emphasized that formal languages do not necessary need to be based on the text. Incredibly efficient way of communication are already mentioned formal languages based on the graphic symbols. Automatic processing of records in such language can be realize by appropriate algorithms, what is another advantage of this languages. Examples of this include automated reasoning systems, which perform symbolic operations basing on simple but incredibly effective methods of processing information.

4 Semantic Knowledge Base

Here, authors would like to introduce a term: *semantic knowledge base*. It is so crucial issue that all discussions on the databases and knowledge bases start and often end with a question: "*what do we mean by the term knowledge base or database?"*. Contrary to appearances, these terms are understood differently depending on the areas of studies of the researchers. The terminological order is essential, not only because of the methodological bases of the science but also for the explanation of solutions proposed by the authors.

4.1 Database

Database is understood as the basic, lowest in the hierarchy of information storage and processing element, from which other levels are built. The definition of the database depends on the domain of the area interested in databases and will not be analyzed herein. Usually the term **database** is understood as stated in Definition 1 or in Definition 2.

Definition 1. *One or more large structured sets of persistent data, usually associated with software to update and query the data. A simple database might be a single file containing many records, each of which contains the same set of fields where each field is a certain fixed width. A database is one component of a database management system.*

Definition 2. *Ordered (having a specific structure) set of information logically connected with each other, which are intended to mirror the fragment of reality. Database should enable the storage of information in a permanent and coherent way that enables access to them (to read, add, delete, modify) anytime in a synchronous way.*

Database should comply with the following requirements:

- to guarantee data integrity,
- to ensure the effective data processing,
- to correctly mirror the relations in the real world represented by a database,
- to protect from the unauthorised access,
- to ensure synchronous access to date to multiple users,
- to make metadata (information about data structure) available.

From the point of view of definition, what is discussed here is not a particular database model: *hierarchical*, *network*, *relational*, *object-oriented*, *association-oriented*. These are only the frames setting the principles of creating categories and relations between them, from which particular bases are constructed. Database consists of a structure. The database structure is defined as a set of elements belonging to the database category together with their mutual relations. For instance, in relational model the structure constitutes a set of tables and relations between this tables. On different levels of structure representation (conceptual, logical, physical), this structure can have different form, degree of precision – generally speaking representation. However, what is really important is the fact that it does not have an established and included in the structure semantics. It means that structure of a database is a structure and semantics describing the meaning of particular elements and relation of these elements is something that should be added in a form of verbal description, and then implemented in a form of application processing data in accordance with a particular structure. Here, it is important to draw attention to the next term, that is database consistency. The term database consistency [19,26,32] is described by the concept verifying whether a database is consistent with a model, that is with a simplified image of reality, which is being represented. The term was mentioned mainly because its relation with the concept of semantics. It is impossible to discuss consistency with a simplified image of the real world and its representation in the form of a structure and limitations without the knowledge of importance of particular elements of this structure. Semantics combines structural elements (syntax or specific structure within a specific syntax) with the outside world, specifically with the understanding (perception) of this world by a modelling person. What is important in the above considerations is that a database in a sense of structure can exist individually and be correct, and the semantics in database is not integrally related with a database itself. In practice, however, it is impossible to image a situation of creating a database structure in isolation of semantic aspects. It is possible only in theory without data, which refers to attributes, combining them together into wholes called in the relational model relations and in object-oriented model classes, creating relations between tables and classes, where all elements function without any relation to anything.

4.2 Knowledge Base

Database can have its structure that can be referred to as a syntax. To this structure it is possible to include semantics, that is determine relations between syntax and what it represents. The next stage and level of development is set of methods, algorithms. Such functions include information about the way of processing data. They can be relatively simple, elementary, e.g. function searching for data that meet specified criteria on the structure and data. However, methods (algorithms) that are being created, in their complexity and level of difficulty are beyond the simple operations on data, e.g. methods of logical inference based on contents of a database. It raises the question that is both philosophical and very practical, from which moment we can or we should say that the database (structure+semantics) consisting of more or less complex function implementing specific tasks can be called a knowledge base. The knowledge base is usually described as a set of information (database) with the ability to interpret those information. Similarly to functions of searching for information, e.g. according to value of the attribute, is hard to call knowledge and to classify the whole system as the knowledge base. However in the case of information system operating on information and consisting of implemented complex algorithms, the inferred situation is no longer unequivocal. Here, the authors wish to emphasize a certain imprecision associated with a blurry statement of "ability of interpretation". It is possible to list a number of examples that are completely analogical to this problem, including Turing's criteria or Chinese Room. In both cases reasonable doubts appear concerning the abilities to "intelligent" processing of information.

In conclusion, the authors would like to propose a certain approach to the issue related to clarification of the concept under the term of knowledge base.

Authors propose the following definition of the knowledge base.

Definition 3. ***Knowledge base*** *(KB) is a system consisting of four elements:*

1. *metastructural database,*
2. *database semantics,*
3. *primary information,*
4. *primary methods of data processing.*

It is proposed to distinguish the following features of KB:

1. *the possibility of freely defining the structure of knowledge,*
2. *the possibility to introduce knowledge,*
3. *the possibility to generate questions to the knowledge base,*
4. *the automatic processing of knowledge.*

The term metastructure should be understood as database structure that is constructed in a way allowing storage of information on the knowledge structure, and only through it the knowledge itself can be stored.

For instance, in a database we can put information of the employee; in a relation model it would be a relation consisting of attributes describing this employee and relationships with other relations, thanks to which it is possible to store information on e.g. his history of employment. It is the data structure. Whereas metastructural database would have the ability to define such type as an employee. This means that it would have the ability to freely, dynamically determine the attributes, which can be possessed by such entity and relations with other entities, both relating to their quantity and other features. Due to that, in a dynamical way, that is the way that does not interfere with a database structure. It is possible to introduce new entities and in the case knowledge base we would refer to terms and their definitions. It is the most important structural element differing the database form the knowledge base. It is, however, not the only difference. Database also can and even should consist of semantics. In the case of knowledge base the semantics is also present, but on the metastructural level. Another important remark is that in theory database can exist without semantics connecting it to the world to be described.

Database does not have to consist of primary data understood as information that is equally important as the structure itself, without which data interpreting and processing would be impossible. In the case of knowledge base, primary data e.g. in the form of primary terms definitions, primary relations (generalization-specialization, part-whole) typical for the knowledge bases, are absolutely essential for the knowledge base to be filled with knowledge.

4.3 Methods of Data Processing

A database does not have to consists of any methods of data processing embedded into its structure. It is a kind of database usage method, the application part that depends on database but is not essential for its existence. The proposed definition, in contrast to unclear, blurry and very general definition describing the knowledge base as a database able to make use of that data, ensures the possibly clear and precise separation of what is a database and what can be called a knowledge base. This is an important issue because there are many publications, in which these terms appear in an unauthorized way, sometimes extremely incorrectly. The example for this include publications, in which non-structuralized sets in the form of text are called knowledge bases. In reality, they are simple repositories that do not even meet the requirements to be treated as knowledge bases. Sets of answers for essential questions are often called knowledge bases, mainly due to the fact that they are sets of data and they refer to some knowledge of a particular field. However, they are not knowledge bases. The classic example are services such as Wikipedia that, as we all know, is an extremely simple technology based on a simple text with references, called hyperlinks. Undoubtedly, it consists of some part of a human knowledge, but it does not make it a knowledge base. It is hard to even imagine that it could be the basis, on which any knowledge base could be built on. The terminological problem can be considered secondary, if not for the methodological issue related to the creation of knowledge bases, and consequently establishing of essential requirements and components of such system.

In the simple definition of the structure of knowledge we should refer to the abolition of restrictions on those that are not a domain of the database. It means that increasing the level of abstraction and entering the metastructure level, where the knowledge structure is possibly optional. The restriction concerns only the metastructure that has the feature of large generality. The possibility to introduce knowledge is analogical to features without data and is based on introducing mechanisms of putting knowledge on the specific structures, and consequently it is related to the mechanisms of structural and semantic correctness control for the particular knowledge base.

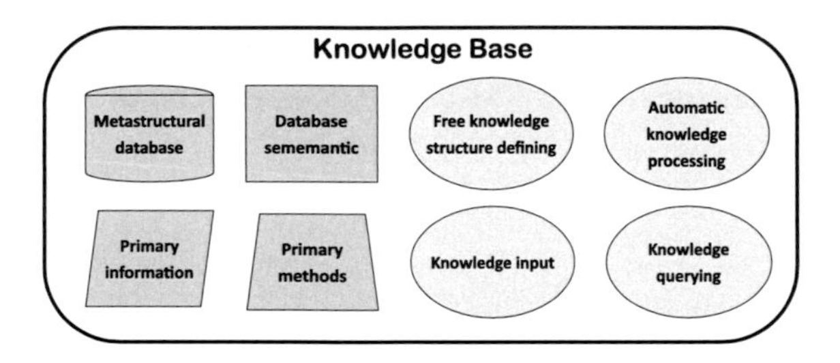

Fig. 6. Knowledge Base construction

The language of questions is also an analogical solution to the language of questions in knowledge bases. It is to ensure the possibility of acquiring knowledge both on the elementary level (such as in databases, where there are mechanisms to read information recorded before) and on the higher level, that is using mechanisms such as acquiring knowledge not only recorded, but also knowledge based on the already acquired knowledge. The example of it can be a mechanism that is able to extract all attributes of a given term taking into account its negative features, but also all features that it has obtained as a result of using inheritance mechanism. It is a simple mechanism, however, it already requires searching the knowledge base and creating more complex semantic response. In databases, such mechanism is not required (Fig. 6).

Automatic processing of knowledge is a definite development of question mechanism. Similarly to the questions mechanism it required the ability to extract data on the basis of information included and more complex information (derived), the mechanism of automatic processing of knowledge relates to much more complex and subtle methods and its purpose is to modify the content of a database. The example of simple knowledge base mechanisms are methods of automatic detection of contradictories and methods of simplifying knowledge, that is conversion form one structure to another in order to optimize efficiency. There are the examples of elementary methods, which does not mean that there are simple to implement, and the level of complexity strictly depends on the level of knowledge base complexity.

4.4 An Example of Knowledge Base

This chapter will present an example of the knowledge base that meets the predefined criteria. At the beginning one more terminological issue needs to be discussed. The developed knowledge base has been named *Semantic Knowledge Base* while in previous chapter it was noted that one of the elements to determine the system as a knowledge base is defined semantics of its database meta structure. Experience shows, however, that very often the term *knowledge base* determines standard data sets, as mention before. This phenomenon is so common that it has become customary, and even though incorrect to fight with it is virtually impossible. Therefore, to distinguish between the knowledge base, understood as different sets of information gathered by people, and the true knowledge base, i.e. one that meets the predefined criteria, the latter will be named *semantic knowledge base*. The term knowledge base is supplemented by "semantic" to clearly highlight the difference between sets of information, that do not have references to the meanings of elements stored in them, and the base, which is focused on the ability to define structures describing these meanings.

As part of research in the field of knowledge engineering *Semantic Knowledge Base* (SKB) system has been developed. It is framework system, i.e. it is not dedicated to a domain-specific applications. This means that it is able to store both, common and specialized knowledge. The common sense knowledge is used to define and to precise more specialized one. SKB is modular and uses logical modules, i.e. there is no strict separation between modules. This is due to the fact that each of the modules cooperates with the other according to the characteristic of the knowledge being processed at the moment.

AODB is a database metamodel designed by authors i.e. for the purpose of knowledge base implementation. It is considered to be the novel database solution, that stems its conceptions from Entity-Relationship approach (E-R) and object-oriented paradigm, representing the database schema as a data and relationship constraint graph whereas the data model is the hypergraph. The syntax and semantics of the metamodel has been deeply elaborated in [15–17]. The two main intensional categories of the metamodel are association and collection. Collection is equivalent to class and is used to store and define objects, and the association is a category of elements that form the relations described by roles. The structure and tasks of the SKB *Structural Module* will be presented as to fulfill the definition of knowledge base in the sense of its metastructure, semantics, as well as, primary information it stores. The description of other modules is not significant since, it would only bring more prove on the same matter, while that description is not the key factor for this paper.

The *Structural Module* consist of two sub-modules: *Ontological Core* (Fig. 7) and *Relationship Module* (Fig. 8). The *Ontological Core* derives its name from the ontology concept, namely the key ontology elements. The most original and structurally situated in the center is an abstract collection CONCEPT. The abstract type is important, since authors assumed that there are no concepts that could not be classified into one of predefined type of concepts. Following collections inherit from an abstract CONCEPT collection: CLASS, INSTANCE, FEATURE, VALUE, VALUESPEC, RELATIONSHIP, SET, which are in turn:

CLASS – represents class in object-oriented model, such as an animal, a person, a feeling.

INSTANCE – represents object in object-oriented model, such as Mickey Mouse, Barrack Obama, i.e. instance that describes a particular element defined by specific class. SKB provides the solution in which it is possible to assign multiple classes to a single instance. Authors do not favor such a solution, however, it was decided to allow the implementation and usage of such a concept.

FEATURE – used to define set of characteristics for a given concept. In the particular case, you can specify the characteristic of the class, what will mean that all instances of this class are described by such features. This mechanism is analogous to attributes assignment for the class in object-oriented model, but much more general, because it can describe the characteristics of any concept, such as relationships, collections, or even other features.

VALUE and VALUESPEC – used to store and define the type of values.

REALATIONSHIP – describes the relationships between concepts through predefined set of attributes as well as any additional features defined by FEATURE. It should be noted that the RELATIONSHIP collection do not represent specific relationships in terms of connections between concepts, but only hold its characteristics. The mechanism of building relations between concepts will be presented later in this paper.

SET – used to describe sets of concepts.

The *Ontologal Core* consists not only of collections, but also associations. These associations, according to AODB model, realize the following relationships.

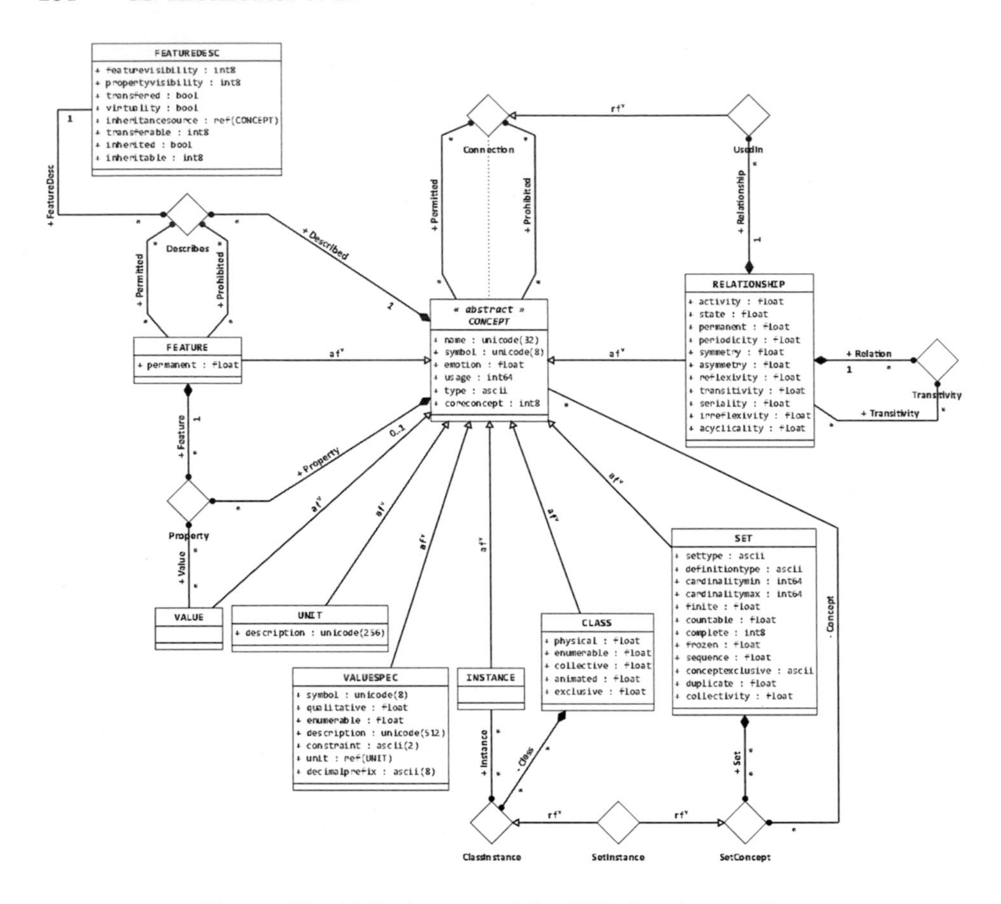

Fig. 7. The AML diagram of the SKB Ontological Core

Property – ties together the collections that are used to describe the properties of a particular concept, i.e. the value of the attribute (FEATURE). The role Property binds the concept with the property, Feature points to the feature being described and Value allows to assign value for the property.

Connection – allows to specify permitted and prohibited connections between concepts. The connection is understood as any possible junction of two concepts, that can be defined on the level of *Ontological Core* or *Relationship Module.*

UsedIn – specialization of Connection used to identify concepts that are permitted or prohibited to define within particular relation.

Describes – used to build lists of features permitted or prohibited to be used for the description of a concept.

ClassInstance – an association in which the instance is ascribed to specific class or classes.

SetConcept, SetInstance – used for building sets of concepts or sets of instances respectively.

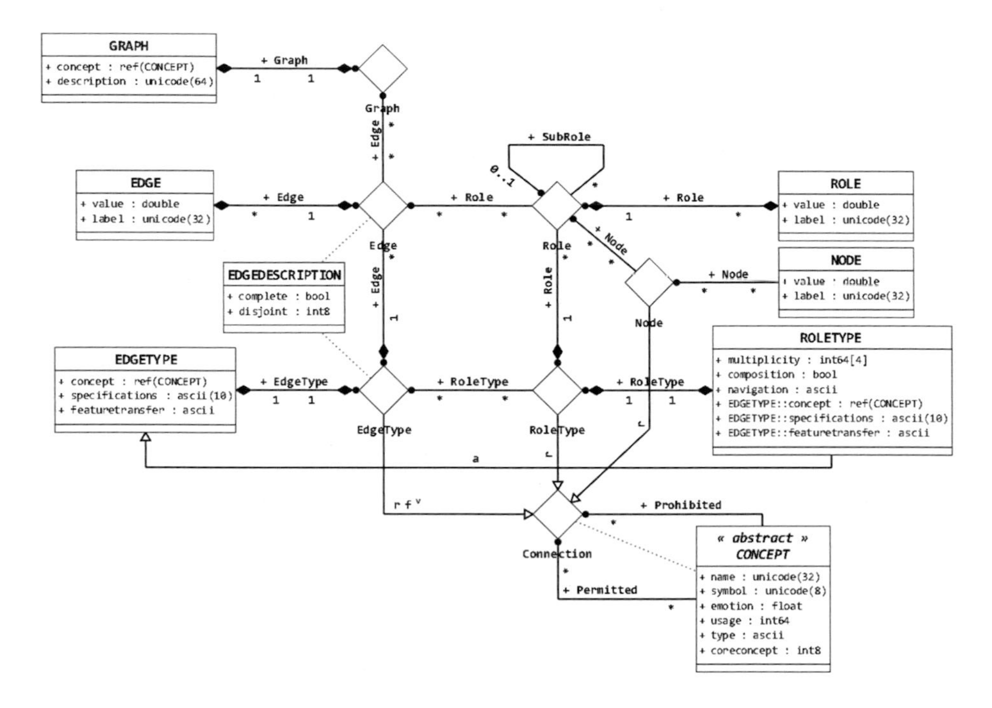

Fig. 8. The structure of *Relationship Module*

Presented *Ontological Core* module uses Gruber's principles regarding ontology design, however it is not a full implementation of the ontology following this approach. This module is used to store the information about concepts and their properties as well as possibilities to establish relations between concepts. In order to create full-fledged ontology, i.e. that, which contains information about the relationships between concepts an *Relationship Module* has been introduced.

Relationship Module consist of two conceptual elements. The first is a system used for building relationship templates, defining type of relation and roles building it. It corresponds to the UML class diagram. The second one is used to determine the specific relationships between concepts defined in Ontological Core. It corresponds to object diagram in UML. Very important issue is that in the process of defining relations and roles only concepts previously defined in *Ontological Core* may be used. Given that *Ontological Core* module is capable of defining any terms, which later might become components of relations in the *Relationship Module*, it should be noted that it gives ability to create any possible relations. Therefore this module is not limited to standard relations, e.g. whole-part, generalization-specialization.

A description of each collection, and association that make up this module is omitted, as it would address the postulate that meeting has already been presented in the description of the *Ontological Core* sub-module. The presented diagram shows the structure of *Relationship Module.* It is worth noticing that

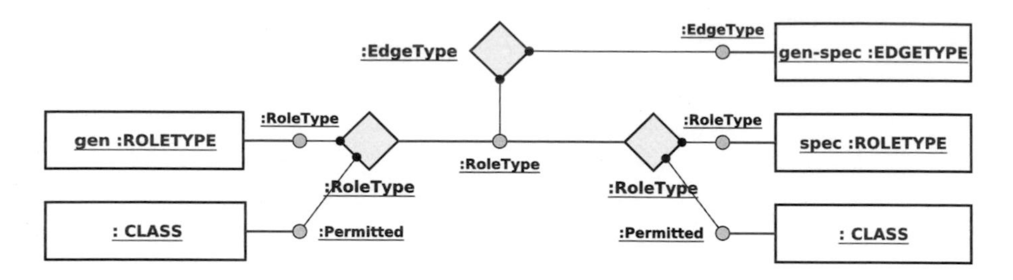

Fig. 9. The definition of generalization-specialization relationship

the relationships are built as hypergraphs, in contrast to classic links derived from object model. The association is treated as an edge, which in turn is made up of roles. The role, in general, is a tree-like structure, that may consist of any number of sub roles. In addition, for each role one can specify the types of terms that may be used by it. This means that at the end of each of the role or sub role may be any number, but at least one node representing the type of the concept. At the stage of defining the role, you can also specify: the multiplicity, the navigation, composition, as well as information regarding inheritance of the attributes of concepts taking part in it.

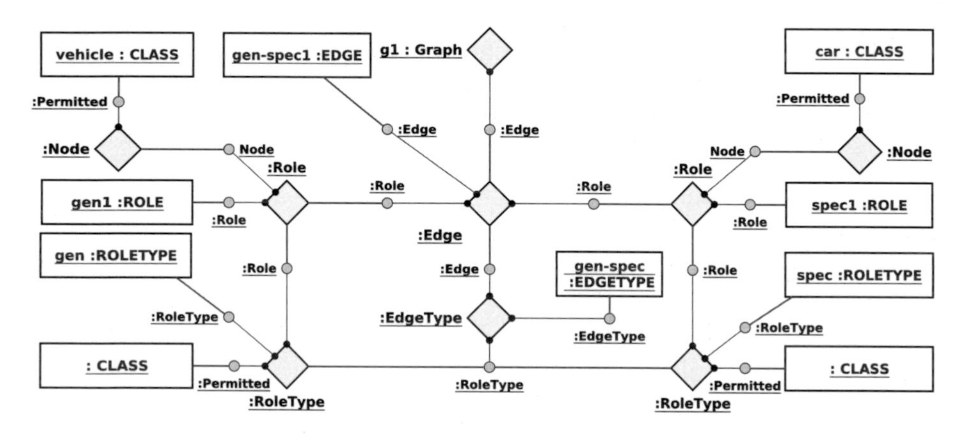

Fig. 10. The example of generalization-specialization relationship implementation

The Fig. 9 shows an example of a simple binary relationship. In the diagram, there is a definition of relationship presented, while in Fig. 10. concrete implementation of this relationship.

5 Summary

The aim of this study was to propose a definition of *Semantic Knowledge Base* in the context of nowadays used terminology and solutions in the field of databases and reality description. Presented methods used for description of reality were not intended to develop discussion in this direction, but rather to briefly present the most common solutions. Each of these may be individually expanded to some extent and interpreted, as well as, implemented in many different ways. The list of solutions is not closed, as there may be a more specific solution or a hybrid one, combining together several of simple methods. Knowledge Bases, depending of on their applications, degree of generality or specialty and implementation may contain different combination of showed methods. This raises the important question about the nature of knowledge base. The paper pointed out that the term Knowledge Base is very often used to name any collection of information resulting from the accumulation of human knowledge. In particular, different repositories, such as sets of answers to frequently asked questions. It is a popular, commercial approach to naming, but has nothing to do with the scientific approach. The source of this is the simple fact that the term database has ceased to be a trendy and catchy carrier, at the moment knowledge is a term often used to catch interest of the recipient. The paper shows that these repositories of information do not have the structure and other attributes that allow them to qualify even as a database. Therefore they might not have nothing to do with the knowledge base. In defining the term of Knowledge Base authors come from the observation that the knowledge is the information along with the possibility of its use. This is an important assumption, however, very general and, unfortunately, very rarely respected. In order to clarify the criteria for determining whether a set of information is a knowledge base, four components and the four properties that the system must meet has been introduced. The authors believe that it is necessary to draw attention to the mass scale abuse of terminology in the field of knowledge engineering. This is very dynamic area and it can be assumed that its development will only accelerate. As a result, it should be a very clear distinction between advanced knowledge base systems solutions and extremely simple repositories of information. For this purpose, authors proposed to supplement the knowledge base term with "semantic" in front. Author is concerned that the term knowledge base has been so widespread that in practice it is impossible to separate legit knowledge base from commercial solutions reaching the marketing terminology.

This article is not supposed to be only a theoretical consideration of the terminology issues, but it also presents a solution namely, *Semantic Knowledge Base* system, developed by the authors. Due to the volume of the studies, the article presents only the main idea, and the most important characteristics of the modules, describing in general terms the most key modules constituting the system core. This system is fully defined by the syntax and semantics in terms of Association-Oriented Database model. AODB is a new model for database modeling, as a key and direct solution, which formed on the basis of the concept of semantic knowledge base. The main idea that joins those studies is the idea of associations being widely used in modeling. A detailed description of the AODB grammar and semantics, due to its size, will be the subject of monographic studies currently being prepared for printing.

References

1. Aagesen, G., Krogstie, J.: BPMN 2.0 for modeling business processes. In: Handbook on Business Process Management 1, pp. 219–250. Springer, Heidelberg (2015)
2. Barnes, W.H.F.: The doctrine of connotation and denotation. Mind **54**, 254–263 (1945)
3. Brown, M.S.: Theaetetus: knowledge as continued learning. J. Hist. Philos. **7**(4), 359–379 (1969)
4. Collins, A.M., Quillian, M.R.: Retrieval time from semantic memory. J. Verbal Learn. Verbal Behav. **8**(2), 240–247 (1969)
5. Dudycz, H.: Approach to the conceptualization of an ontology of an early warning system. In: Information Systems in Management XI, Data Bases, Distant Learning, and Web Solutions Technologies, pp. 29–39 (2011)
6. Duhl, J., Damon, C.: A performance comparison of object and relational databases using the sun benchmark. In: ACM SIGPLAN Notices, vol. 23, pp. 153–163. ACM (1988)
7. Feng, S., Bose, R., Choi, Y.: Learning general connotation of words using graph-based algorithms. In: Proceedings of the Conference on Empirical Methods in Natural Language Processing, pp. 1092–1103. Association for Computational Linguistics (2011)
8. French, R.M.: The chinese room: Just say "no!" In: Proceedings of the Cognitive Science Society, vol. 1 (2000)
9. Gruber, T.R.: A translation approach to portable ontology specifications. Knowl. Acquisition **5**(2), 199–220 (1993)
10. Hao, C.: Research on knowledge model for ontology-based knowledge base. In: 2011 International Conference on Business Computing and Global Informatization (BCGIN), pp. 397–399. IEEE (2011)
11. Hendrix, G.G.: Encoding knowledge in partitioned networks. In: Associative Networks: Representation and Use of Knowledge by Computers, pp. 51–92 (1979)
12. Joyce, D.: An identification and investigation of software design guidelines for using encapsulation units. J. Syst. Softw. **7**(4), 287–295 (1987)
13. Kalantari, R., Bryant, C.: Comparing the performance of object and object relational database systems on objects of varying complexity. In: Data Security and Security Data, pp. 72–83 (2012)
14. Korzynska, A., Zdunczuk, M.: Clustering as a method of image simplification. Inf. Technol. Biomed. **47**, 345 (2008)
15. Krótkiewicz, M.: Asocjacyjny metamodel baz danych. Definicja formalna oraz analiza porównawcza metamodeli baz danych (eng. Association-Oriented Database Metamodel). No. z. 444 in Studia i Monografie, Oficyna Wydawnicza Politechniki Opolskiej, Opole (2016)
16. Krótkiewicz, M.: Association-oriented database model - n-ary associations. Int. J. Softw. Eng. Knowl. Eng. **27**, 281 (2017)
17. Krótkiewicz, M., Wojtkiewicz, K., Jodłowiec, M., Pokuta, W.: Semantic knowledge base: quantiers and multiplicity in extended semantic networks module. In: Knowledge Engineering and Semantic Web: 7th International Conference, KESW 2016, Prague, Czech Republic, 21–23 September 2016, Proceedings. Springer, Cham (2016)
18. Lange, K.J.: Complexity and structure in formal language theory. Fundam. Inf. **25**(3, 4), 327–352 (1996)

19. Lin, K.J.: Consistency issues in real-time database systems. In: Proceedings of the Twenty-Second Annual Hawaii International Conference on System Sciences, 1989. Vol. II: Software Track, vol. 2, pp. 654–661. IEEE (1989)
20. Macewen, G.H., Martin, T.P.: Abstraction hierarchies in top-down design. J. Syst. Softw. **2**(3), 213–224 (1981)
21. OMG: Unified Modeling Language^{TM} (UML®) Version 2.5 (2013). http://www.omg.org/spec/UML/2.5/www.omg.org/spec/UML/2.5/Beta2/PDF/
22. Przepiórkowski, A.: Slavonic information extraction and partial parsing. In: Proceedings of the Workshop on Balto-Slavonic Natural Language Processing: Information Extraction and Enabling Technologies, pp. 1–10. Association for Computational Linguistics (2007)
23. Przepiórkowski, A., Górski, R.L., Lewandowska-Tomaszyk, B., Lazinski, M.: Towards the national corpus of polish. In: LREC (2008)
24. Przepiórkowski, A., Marcińczuk, M., Degórski, Ł.: Dealing with small, noisy and imbalanced data. In: Text, Speech and Dialogue, pp. 169–176. Springer, Heidelberg (2008)
25. Schärli, N., Black, A.P., Ducasse, S.: Object-oriented encapsulation for dynamically typed languages. In: ACM SIGPLAN Notices, vol. 39, pp. 130–149. ACM (2004)
26. Seligman, L.J., Kerschberg, L.: Knowledge-base/database consistency in a federated multidatabase environment. In: Proceedings of the Third International Workshop on Research Issues in Data Engineering: Interoperability in Multidatabase Systems, RIDE-IMS 1993, pp. 18–25. IEEE (1993)
27. Soutou, C.: Modeling relationships in object-relational databases. Data Knowl. Eng. **36**(1), 79–107 (2001)
28. Stroustrup, B.: What is object-oriented programming? IEEE Softw. **5**(3), 10–20 (1988)
29. Su, H., Bouridane, A., Crookes, D.: Scale adaptive complexity measure of 2D shapes. In: 18th International Conference on Pattern Recognition, ICPR 2006, vol. 2, pp. 134–137. IEEE (2006)
30. Voigt, J., Irwin, W., Churcher, N.: Class encapsulation and object encapsulation: an empirical study (2010)
31. Wislicki, J., Kuliberda, K., Adamus, R., Subieta, K.: Relational to object-oriented database wrapper solution in the data grid architecture with query optimisation issues. Int. J. Bus. Process Integ. Manag. **2**(1), 17–25 (2007)
32. Zhangbing, L., Wujiang, C.: A new algorithm for data consistency based on primary copy data queue control in distributed database. In: 2011 IEEE 3rd International Conference on Communication Software and Networks (ICCSN), pp. 207–210. IEEE (2011)

Author Index

W. P. Hunek and S. Paszkiel (Eds.): BCI 2018, AISC 720, p. 241, 2018.
https://doi.org/10.1007/978-3-319-75025-5

Printed in the United States
By Bookmasters